Patterns and Figures

Britannica® Mathematics in Context

Algebra

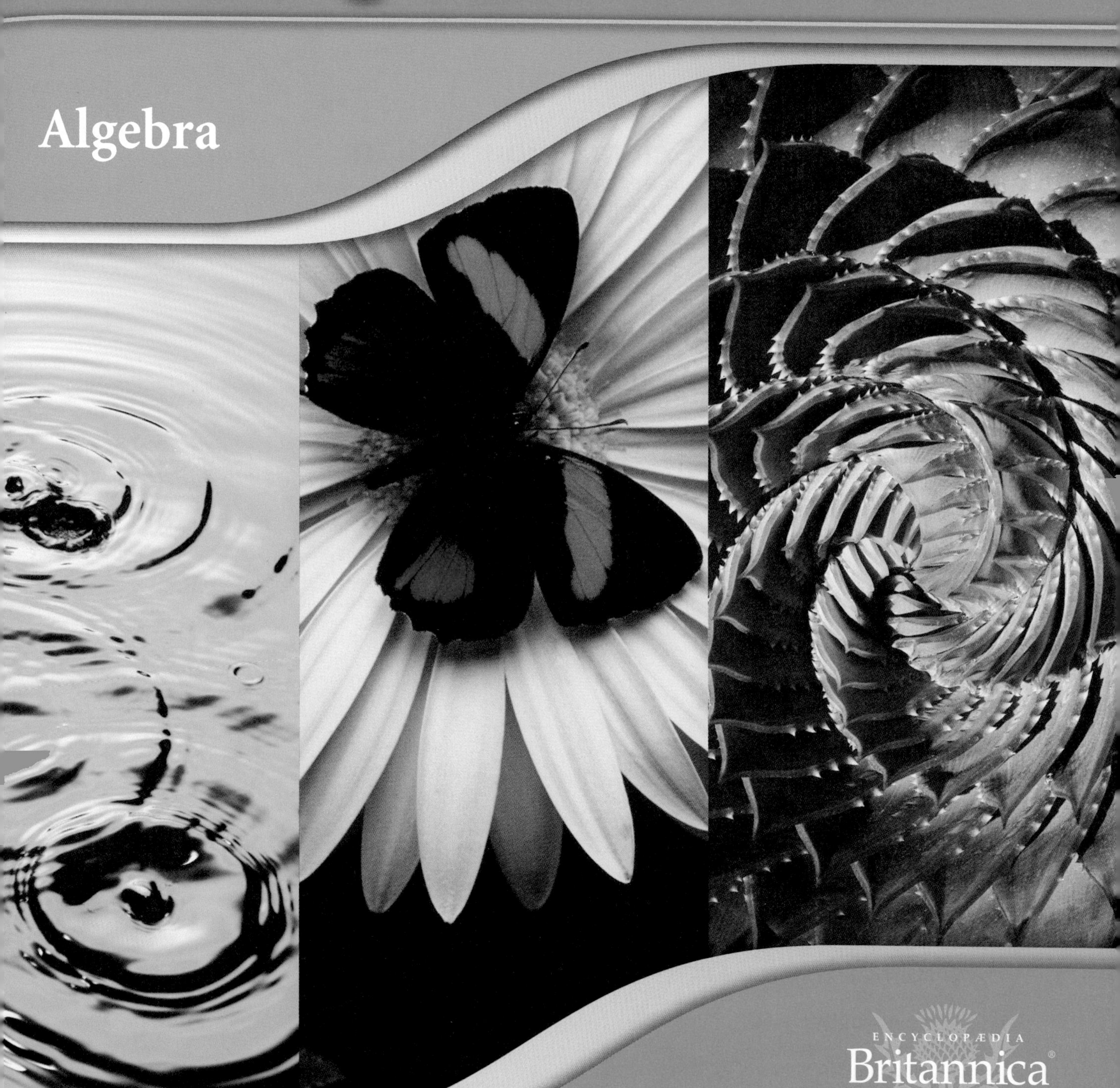

ENCYCLOPÆDIA Britannica®

Mathematics in Context is a comprehensive curriculum for the middle grades. It was developed in 1991 through 1997 in collaboration with the Wisconsin Center for Education Research, School of Education, University of Wisconsin-Madison and the Freudenthal Institute at the University of Utrecht, The Netherlands, with the support of the National Science Foundation Grant No. 9054928.

The revision of the curriculum was carried out in 2003 through 2005, with the support of the National Science Foundation Grant No. ESI 0137414.

National Science Foundation

Opinions expressed are those of the authors
and not necessarily those of the Foundation.

Kindt, M., Roodhardt, A., Wijers, M., Dekker, T., Spence, M. S., Simon, A. N., Pligge, M. A., & Burrill, G. (2010). Patterns and figures. In Wisconsin Center for Education Research & Freudenthal Institute (Eds.), *Mathematics in context.* Chicago: Encyclopædia Britannica, Inc.

International Standard Book Number 978-1-59339-962-7

Printed in the United States of America

1 2 3 4 5 C 13 12 11 10

The *Mathematics in Context* Development Team

Development 1991–1997

The initial version of *Patterns and Figures* was developed by Martin Kindt and Anton Roodhardt. It was adapted for use in American schools by Mary S. Spence, Aaron N. Simon, and Margaret A. Pligge.

Wisconsin Center for Education Research Staff

Thomas A. Romberg
Director

Joan Daniels Pedro
Assistant to the Director

Gail Burrill
Coordinator

Margaret R. Meyer
Coordinator

Project Staff

Jonathan Brendefur
Laura Brinker
James Browne
Jack Burrill
Rose Byrd
Peter Christiansen
Barbara Clarke
Doug Clarke
Beth R. Cole
Fae Dremock
Mary Ann Fix
Sherian Foster
James A, Middleton
Jasmina Milinkovic
Margaret A. Pligge
Mary C. Shafer
Julia A. Shew
Aaron N. Simon
Marvin Smith
Stephanie Z. Smith
Mary S. Spence

Freudenthal Institute Staff

Jan de Lange
Director

Els Feijs
Coordinator

Martin van Reeuwijk
Coordinator

Mieke Abels
Nina Boswinkel
Frans van Galen
Koeno Gravemeijer
Marja van den Heuvel-Panhuizen
Jan Auke de Jong
Vincent Jonker
Ronald Keijzer
Martin Kindt
Jansie Niehaus
Nanda Querelle
Anton Roodhardt
Leen Streefland
Adri Treffers
Monica Wijers
Astrid de Wild

Revision 2003–2005

The revised version of *Patterns and Figures* was developed by Monica Wijers and Truus Dekker. It was adapted for use in American schools by Gail Burrill.

Wisconsin Center for Education Research Staff

Thomas A. Romberg
Director

David C. Webb
Coordinator

Gail Burrill
Editorial Coordinator

Margaret A. Pligge
Editorial Coordinator

Project Staff

Sarah Ailts
Beth R. Cole
Erin Hazlett
Teri Hedges
Karen Hoiberg
Carrie Johnson
Jean Krusi
Elaine McGrath
Margaret R. Meyer
Anne Park
Bryna Rappaport
Kathleen A. Steele
Ana C. Stephens
Candace Ulmer
Jill Vettrus

Freudenthal Institute Staff

Jan de Lange
Director

Truus Dekker
Coordinator

Mieke Abels
Content Coordinator

Monica Wijers
Content Coordinator

Arthur Bakker
Peter Boon
Els Feijs
Dédé de Haan
Martin Kindt
Nathalie Kuijpers
Huub Nilwik
Sonia Palha
Nanda Querelle
Martin van Reeuwijk

Cover photo credits: (all) © Corbis

Illustrations
7 Holly Cooper-Olds; **12** Steve Kapusta/© Encyclopædia Britannica, Inc.; **15, 17** James Alexander; **29** Holly Cooper-Olds; **34** James Alexander

Photographs
1 (left to right) © Pixtal; Brand X Pictures/Alamy; © Corbis; **6** PhotoDisc/ Getty Images; **20** Alvaro Ortiz/HRW Photo; **24** Victoria Smith/HRW; **35** BananaStock/Alamy

Contents

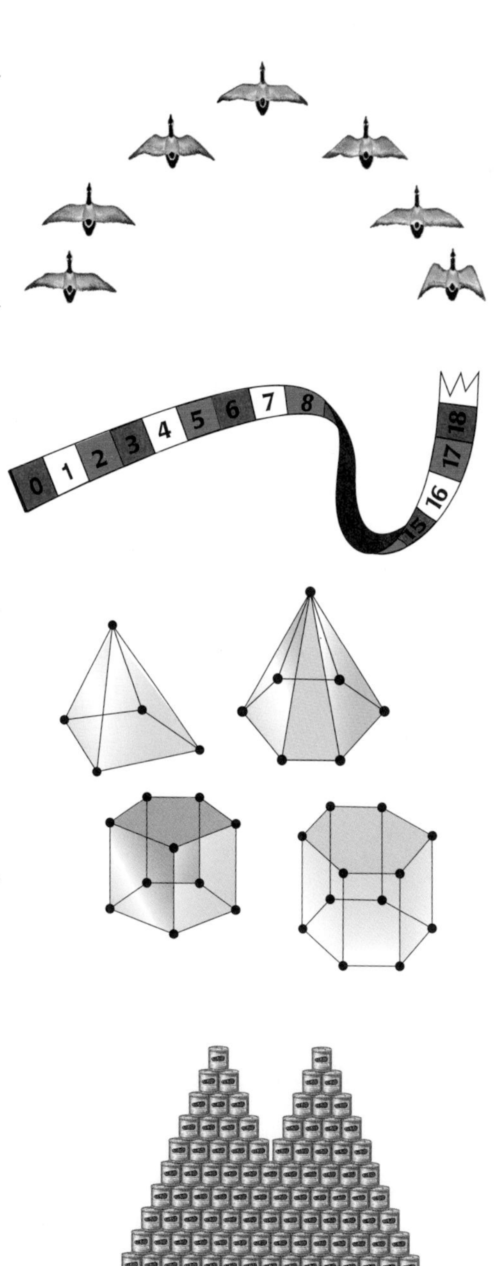

Dear Student,

Welcome to the unit *Patterns and Figures*. In this unit, you will identify patterns in numbers and shapes and describe those patterns using words, diagrams, and formulas.

You have already seen many patterns in mathematics. For patterns with certain characteristics, you will learn rules and formulas to help you describe them. Some of the patterns are described by using geometric figures, and others are described by a mathematical relationship.

Here are two patterns. One is a pattern of dots, and the other is a pattern of geometric shapes.

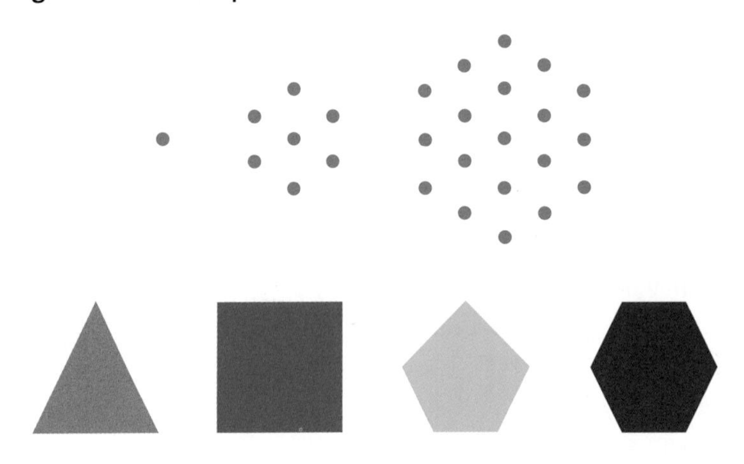

Can you describe the dot pattern? Where do you think the pattern of shapes ends?

As you investigate the *Patterns and Figures* unit, remember that patterns exist in many places—almost anywhere you look! The skills you develop in looking for and describing patterns will always help you, both inside and outside your math classroom.

Sincerely,

The Mathematics in Context Development Team

Patterns

Number Strips

Patterns are at the heart of mathematics, and you can find patterns by looking at shapes, numbers, and many other things. In this unit, you will discover and explore patterns and describe them with numbers and formulas.

Below, numbers starting with 0 are shown on a paper strip. The strip has alternating red and white colors.

1. Notice that the right end of the strip looks different from the left end. What do you think that indicates?

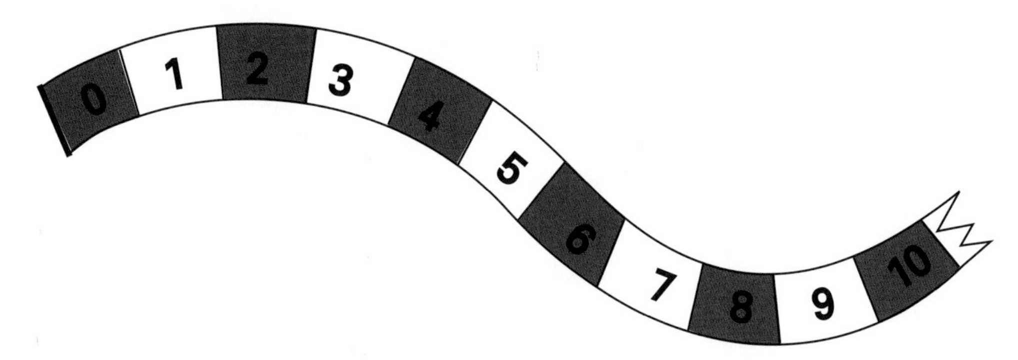

2. a. What do the white numbers have in common?

   b. Think of a large number not shown on the strip. How can you tell the color for your number?

Here is a different strip made with the repeating pattern red – white – blue — red — white — blue.

Any list of numbers that goes on forever is called a **sequence**.

3. How can you figure out the color in the red-white-blue sequence for 253,679?

One way to "see" a pattern is to use dots to represent numbers. For example, the red numbers from the red and white strip on page 1 can be drawn like this:

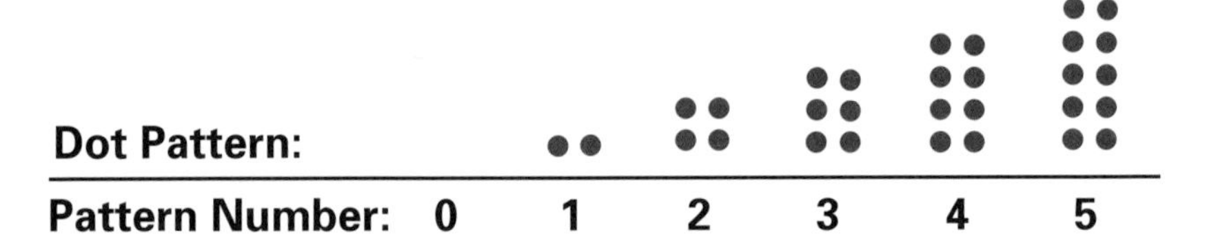

Below each dot pattern is a pattern number. The pattern number tells you where you are in a sequence. (Notice that the pattern number starts with 0, and there are no dots for pattern number 0.)

Pattern number 1 shows two dots, pattern number 2 shows four dots, and so on, assuming that the pattern continues building dots in the same way.

4. a. Look at the dot pattern for the red numbers. When the pattern number is 37, how many dots are there?

b. Someone came up with the formula $R = 2n$ for the red numbers. What do you think R and n stand for?

c. Does the formula work? Explain your answer.

You can represent the white numbers from the red-white strip on page 1 in their own pattern: 1, 3, 5, ...

These numbers can be represented using a different dot pattern as shown below.

Dot Pattern:	●	● ● ●	● ● ● ● ●	● ● ● ● ● ● ●	● ● ● ● ● ● ● ● ●	● ● ● ● ● ● ● ● ● ● ●
Pattern Number:	**0**	**1**	**2**	**3**	**4**	**5**

5. a. Now look at the pattern for the white numbers. How many dots are in pattern number 50?

b. Write a formula for the white numbers.

Rule: ***"If you add two odd numbers, you get an even number."***

6. a. Use dots to explain the rule above.

b. Make up some other rules like the one above, and use dots to explain them.

The sequence of even numbers {0, 2, 4, 6, 8 ...} can be described by the formula:

START number = 0
NEXT even number = CURRENT even number + 2

You may have seen these "NEXT-CURRENT" formulas in previous *Mathematics in Context* units. They are more formally called **recursive formulas.**

7. a. Write a NEXT-CURRENT formula for the sequence of odd numbers

{1, 3, 5, 7}.

b. Compare the formulas for even and odd numbers assuming that the pattern continues building dots in the same way. What is the same and what is different?

A formula such as those you found above for even and odd numbers is called a **direct formula.**

8. Why do you think these are called direct formulas?

Look again at the red–white–blue sequence from page 2.

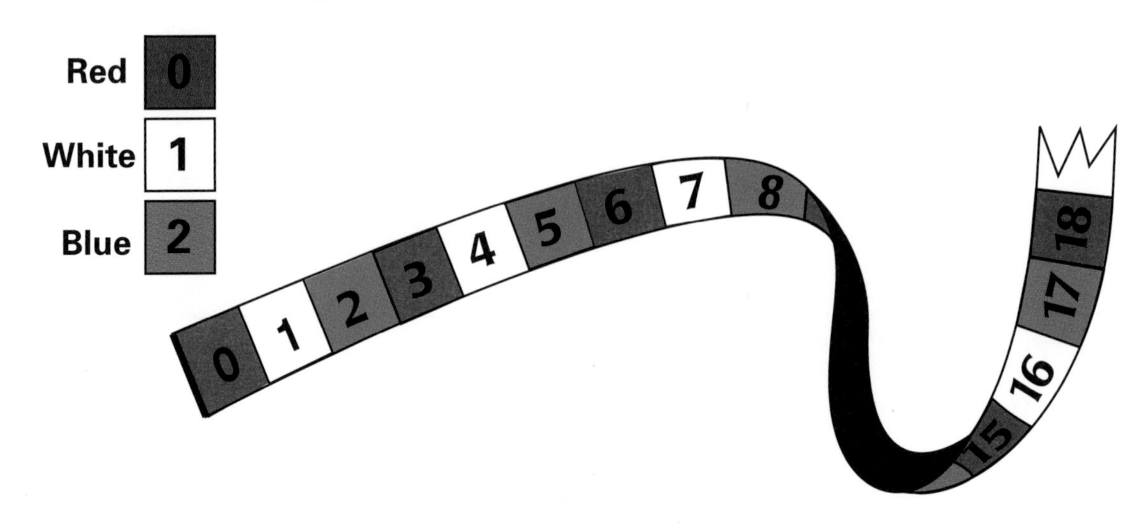

9. a. Represent the red, white, and blue numbers using dot patterns similar to the dot patterns shown on pages 2 and 3.

b. Write a NEXT-CURRENT formula and a direct formula for the sequence of red numbers. State where your sequence begins in both cases.

c. Do the same for the sequences of white numbers and blue numbers.

d. If you add a white number to a blue number, do you always get a red number? Use dots to explain your answer.

e. Copy the chart in your notebook and complete it for all combinations of colors.

+	R	W	B
R			
W			
B			

V- and W-Formations

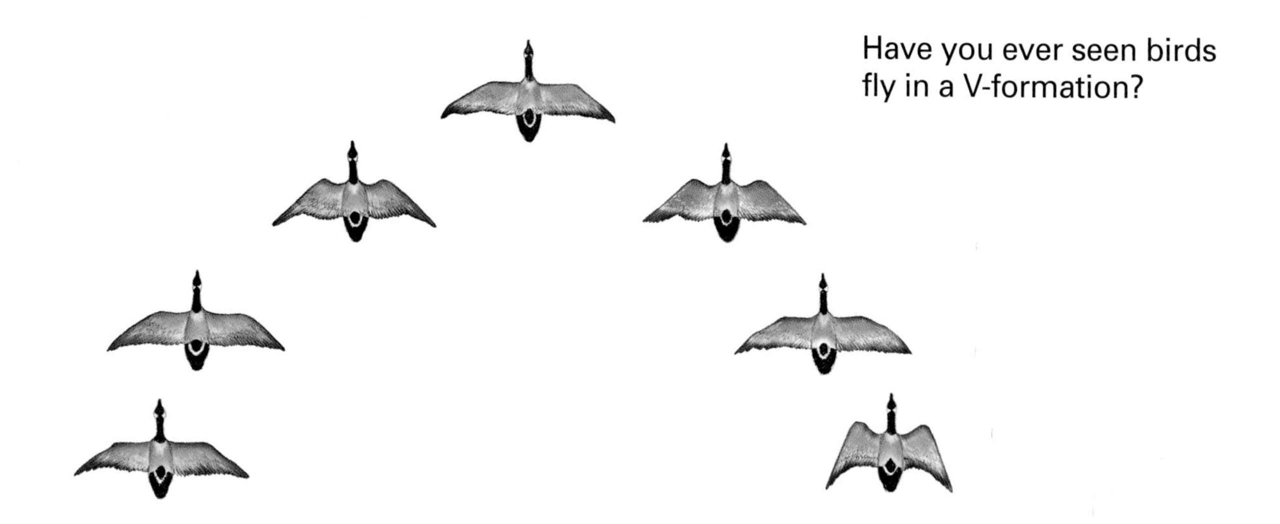

Have you ever seen birds fly in a V-formation?

You can make a sequence of V-patterns using dots. The first four are shown below.

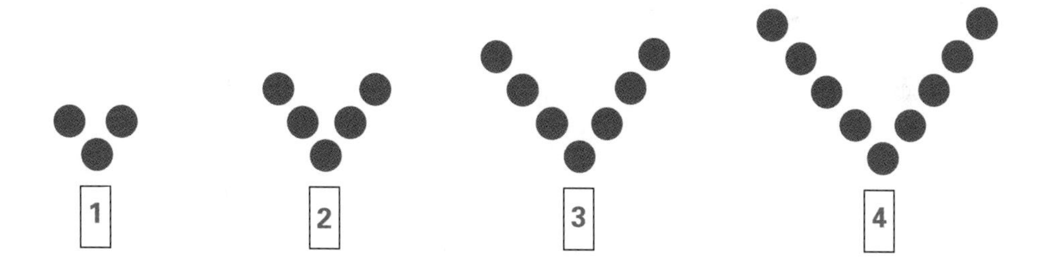

Here is a V-number that uses 17 dots.

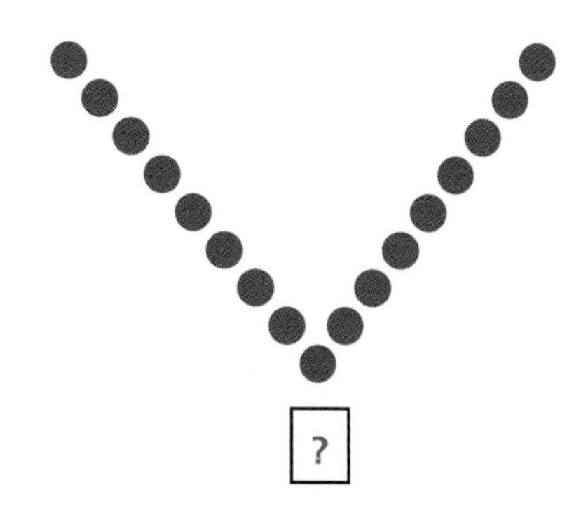

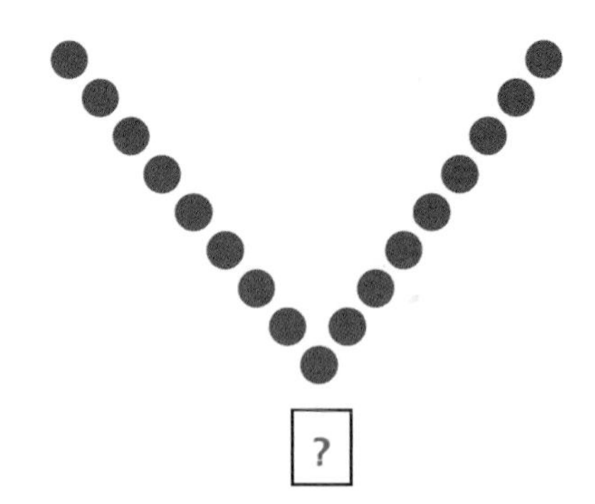

10. a. What is the pattern number of this drawing?

    b. How many dots are in pattern number 85?

    c. Is it possible to make a V-pattern with 35,778 dots? Explain why or why not.

    d. Write a direct formula to describe the number of dots in any V-pattern. $V = ?$

The letter n is usually used in direct formulas to denote pattern numbers.

For some patterns n starts with 0 and sometimes with 1. It may start with other numbers as well.

11. If you haven't already done so, write your V-pattern formula using the letter n.

Squadrons of airplanes sometimes fly in a W-formation.

Look at the following sequence of W-patterns.

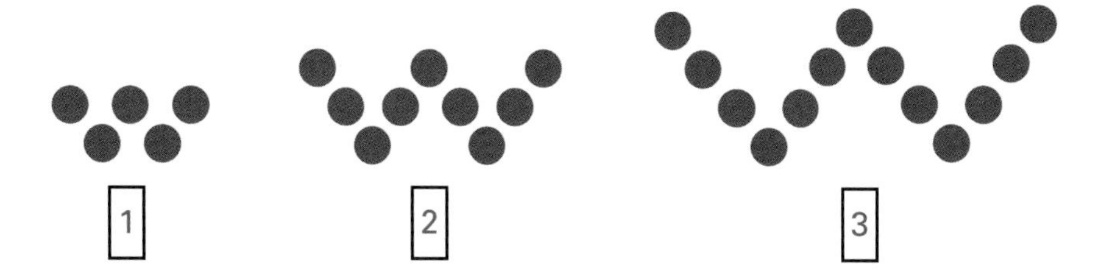

12. a. Copy and complete this chart for the W-patterns.

Number of Dots	5					
Pattern Number	1	2	3	4	5	6

b. Write a NEXT-CURRENT formula for the number of dots in the W-pattern sequence.

c. How many dots are in pattern number 16?

d. Find a direct formula to describe the number of dots in any W-pattern. Then use your formula to find the number of dots in pattern number 25.

Jessie compared the W-patterns to the V-patterns. She said, "W is double V minus one."

13. a. What did Jessie mean? Use dot patterns to explain her statement.

b. Explain Jessie's statement using direct formulas for $W = \ldots$ and $V = \ldots$

The Williams Pie Company wants to display a big **"W"** with orange light bulbs on a billboard. They order 200 light bulbs.

14. If they place the light bulbs in a W-pattern, how many bulbs would there be in the largest W they can make?

A Patterns

Summary

Number strips and dot patterns can illustrate sequences of numbers. You can use formulas to describe sequences and to find numbers later in the sequence. Here is an example where the dots continue to build in the same way.

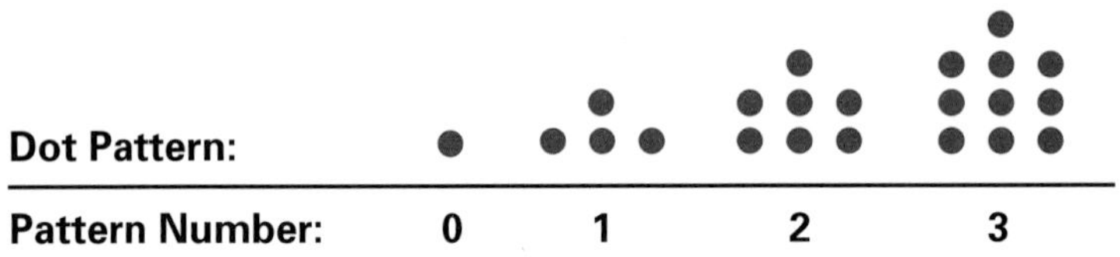

A NEXT-CURRENT formula or *recursive* formula has two parts: a start value and a rule for finding each "next" value. A recursive formula for the number of dots in the dot pattern above is:

START number = 1
NEXT number = CURRENT number + 3

A *direct formula* uses n to indicate the pattern number. If D stands for the number of dots, a direct formula for the pattern above is:

$D = 3n + 1$, where n starts with 0.

Check Your Work

1. James wrote the direct formula $D = 1 + n + n + n$ for the dot pattern in the Summary. Show whether or not James's formula is correct.

Look again at the dot pattern of page 3.

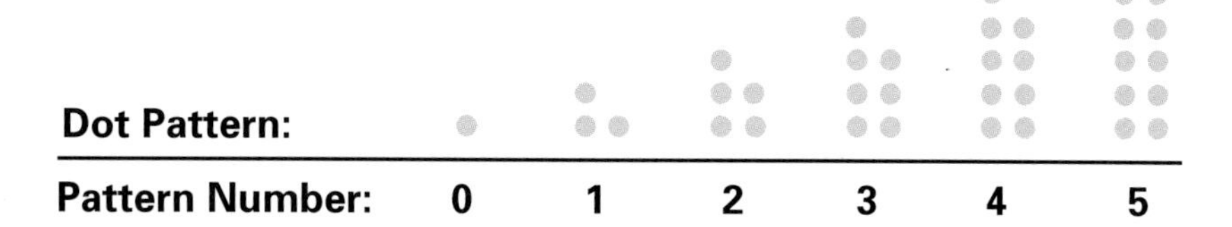

David came up with the formula $W = 2n + 1$ for this pattern.
Cindy found the direct formula $W = 2(n + 1) - 1$

2. **a.** Are both formulas correct? Why or why not?

   b. How would David's formula change if the first pattern number is 1 instead of 0?

3. **a.** Make a sequence of dot patterns for the direct formula $D = 5n + 2$, where n starts at 0.

   b. Write a NEXT-CURRENT formula for the sequence.

4. **a.** Make up your own dot-pattern sequence.

   b. Write a direct formula and a NEXT-CURRENT formula for your sequence.

5. **Reflect** What is one advantage and one disadvantage of a recursive formula compared to a direct formula?

For Further Reflection

Can every sequence of numbers be represented by a dot pattern? Why or why not?

B Sequences

Constant Increase/Decrease

A sequence that has a constant increase or decrease is called an **arithmetic sequence.** Here is an example of an arithmetic sequence. The jagged right end indicates that the strip continues forever.

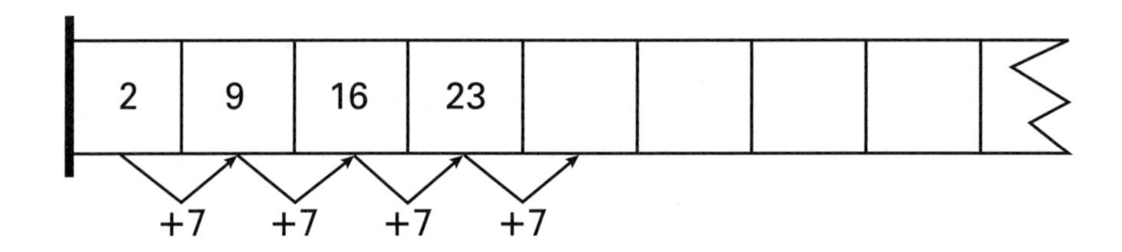

1. **a.** Write four more numbers as they appear on the strip.

   b. Will the number 100 be on the number strip? How do you know?

   c. How about the number 200?

   d. Write a large number that will never appear on the strip. How do you know for sure that it will never appear?

Jorge came up with the expression $2 + 7n$ for the number strip shown on this page.

2. **a.** Where does n start in Jorge's expression?

   b. Use the expression to find the next three numbers on the strip.

Here is a number strip showing a different arithmetic sequence.

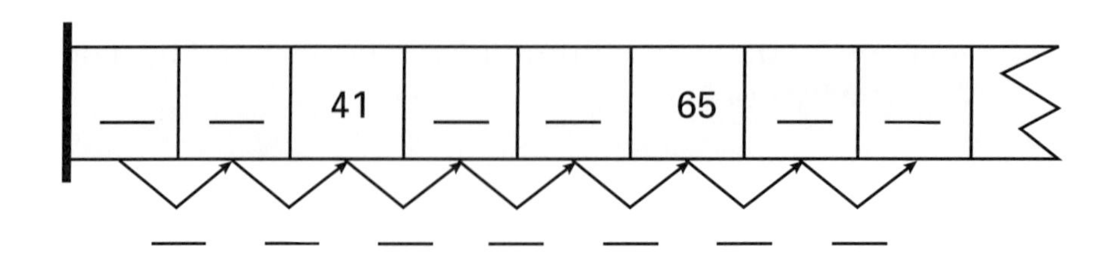

3. **a.** Find the missing numbers on the strip.

   b. Write a recursive formula for this number strip.

   c. Write an expression for this number strip. Let n start at zero.

Instead of adding a number at each step, some arithmetic sequences subtract a number at each step.

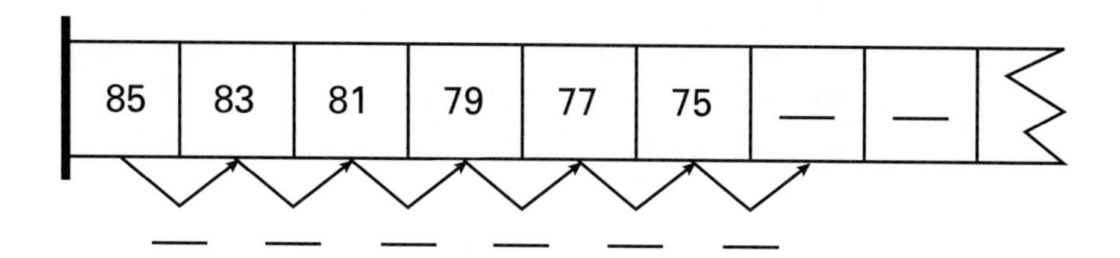

4. a. What is the decrease in this sequence?

b. Write an expression for this number strip. Let n start at 0.

c. Explain the difference between a (direct) formula and an expression.

d. How many steps does it take to get to the first negative number in this strip?

5. Find an expression for the arithmetic sequence in the number strip drawn below. Let n start at 0 in your expression. You may copy the strip in your notebook and fill in the missing numbers first if you want to.

$3\frac{3}{4}$	$3\frac{1}{2}$		3		$2\frac{1}{2}$	$2\frac{1}{4}$	

6. Write the first five numbers in each of the sequences described by the following expressions. For each sequence, n starts at zero.

a. $4 - 3n$

b. $2\frac{1}{2} + \frac{1}{2}n$

c. $5n - 10$

d. $12n$

Many sequences are not arithmetic. Some involve multiplication and division.

7. a. Design a sequence that has a pattern or regularity but is not an arithmetic sequence.

b. Describe the regularity in your sequence.

Adding and Subtracting Strips

Larry's favorite number strip is the sequence of odd numbers. He decides to add his number strip of odd numbers to the strip of even numbers.

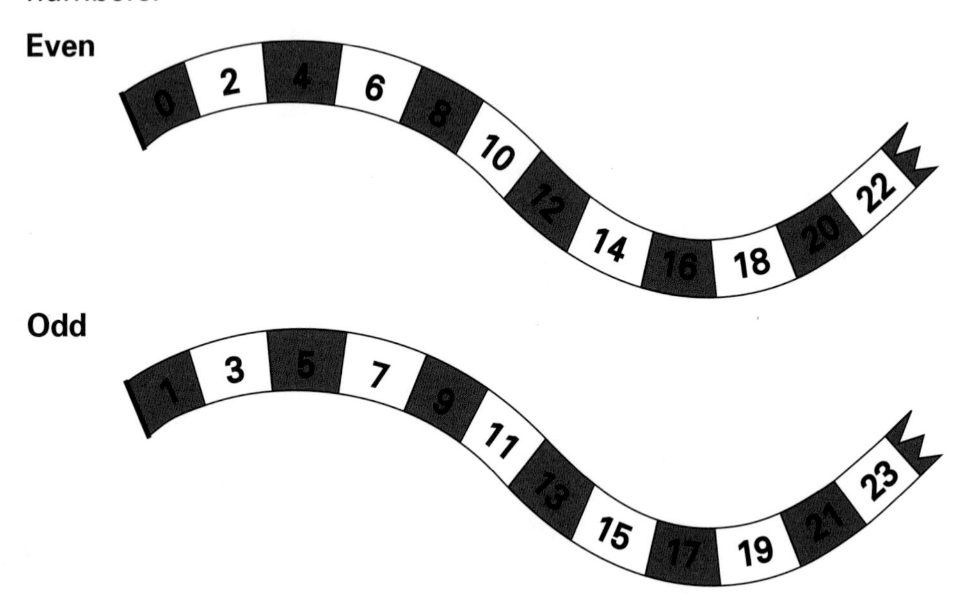

Here are the first three numbers of the resulting sequence:

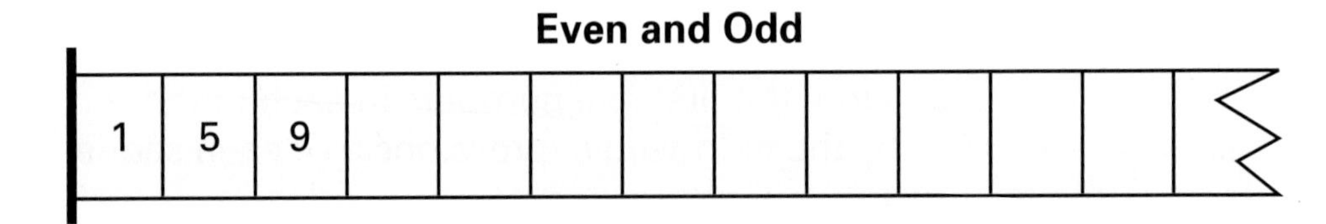

8. Copy the number strip above and find the next nine numbers in the new sequence.

There is a connection between the W-patterns shown and the new number strip.

9. What is that connection?

10. a. Write an expression for each of the three numbers strips in problem 8. Let n start at zero.

b. How can you use your expressions to check that the third sequence is the sum of the other two?

Compare these three number strips:

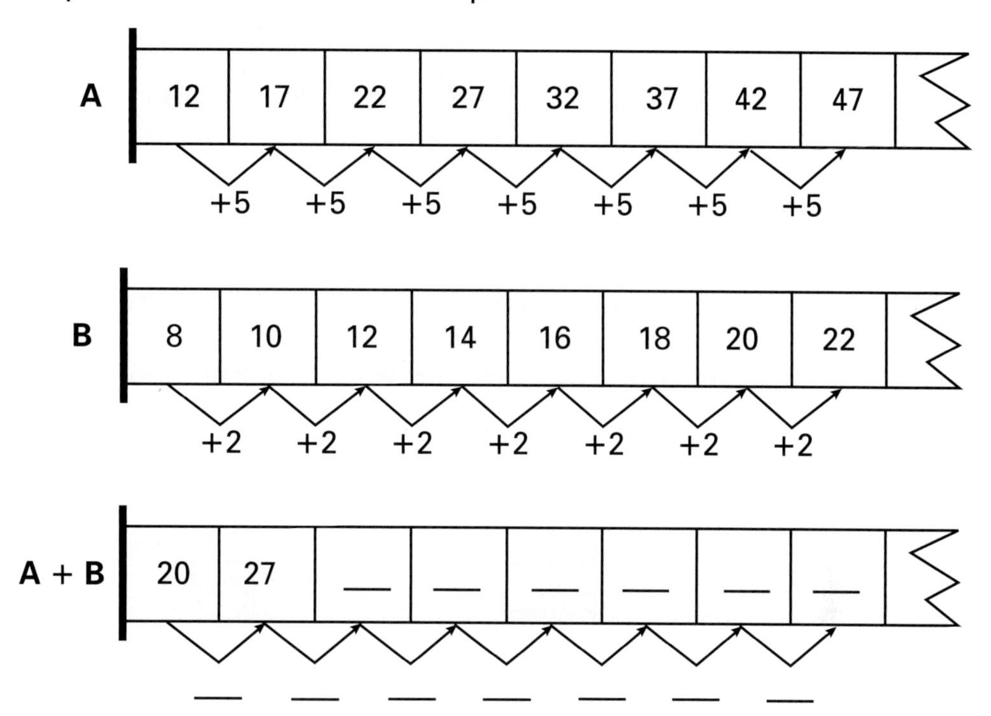

Number strips A and B are arithmetic sequences. The entries in the third strip, A + B, are formed by adding the numbers that appear in the same positions on strips A and B.

11. a. Without filling in the numbers for the strip A + B, you can show that it must be an arithmetic sequence. Explain how.

b. Write expressions for the number strips A, B, and A + B.

c. Make a number strip for A – B. Do the numbers form an arithmetic sequence? Explain why or why not.

d. What expression corresponds to the number strip A – B?

Strips C and D are two other number strips. The expression for strip C is $6 + 3n$, and the expression for strip D is $4 + 5n$. For both strips, n starts at zero.

Jim wants to make an expression for the strip C + D. First he makes the strips C and D. Then he adds the numbers on the two strips to get the numbers on strip C + D.

Then he makes the expression for the strip C + D.

12. Show the three steps in Jim's solution.

Gail thinks she knows a shorter way to come up with the expression.

13. a. What might Gail have in mind?

b. What is the expression for the number strip C – D?

14. Copy the three strips into your notebook. Fill in the missing numbers and write the expression for each number sequence.

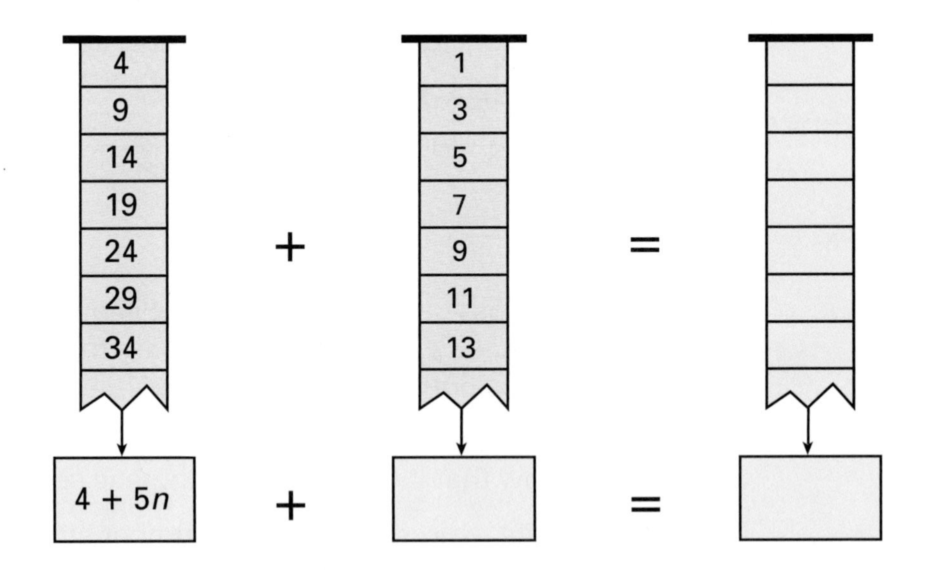

15. Copy the three strips shown below. Fill in the missing numbers and write an expression for each number sequence.

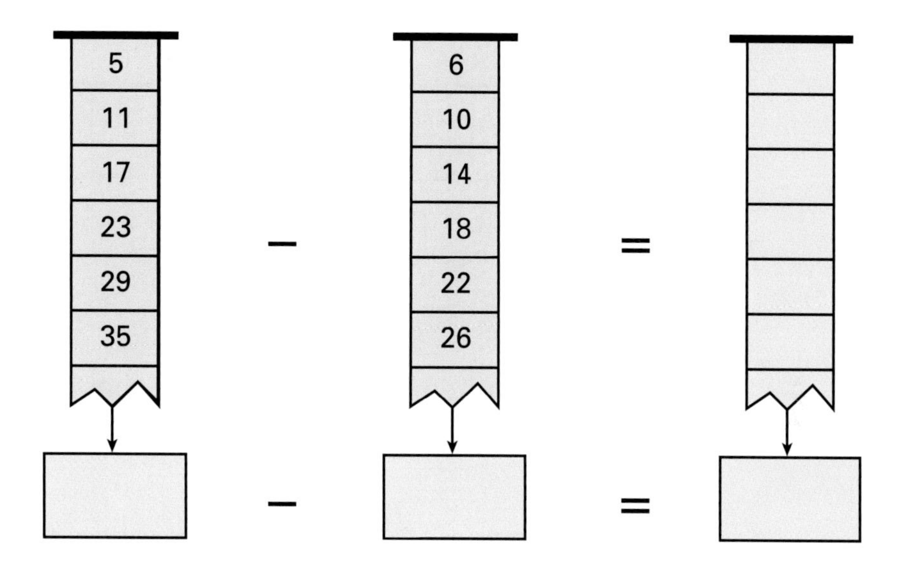

16. a. Write an expression for the sum of $17 + 5n$ and $13 - 7n$.

b. Write an expression for the difference of $17 + 5n$ and $13 - 7n$. Use number strips to show why your answer makes sense.

Pyramids

Billy is a glass artist who makes geometric shapes out of glass. Here is a sequence of his pyramids.

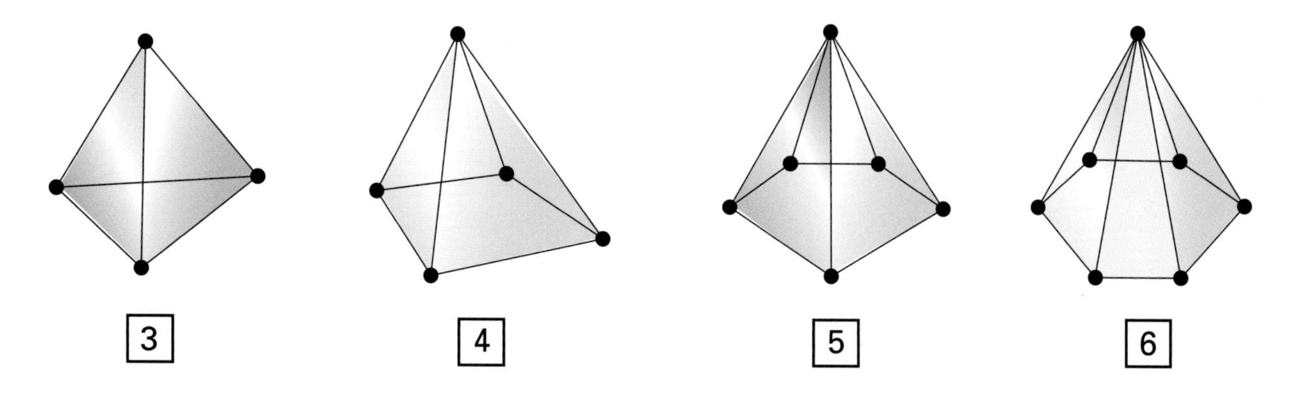

17. Explain the numbers below the pyramids.

This number strip represents the number of vertices (V) for the sequence of pyramids. A **vertex** is the intersection of the edges of the pyramid.

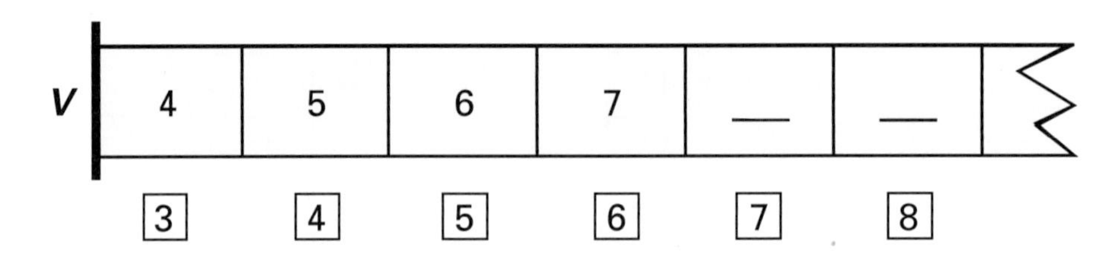

18. Find a formula for number strip V that relates V (the number of vertices) to n (the numbers below the pyramids). Where does n start?

19. **a.** Make number strips for the numbers of edges (E) and the numbers of faces (F) for the sequence of pyramids.

b. Write formulas for number strips E and F.

c. Combine number strips V, E, and F into a new number strip whose formula shows $V - E + F$.

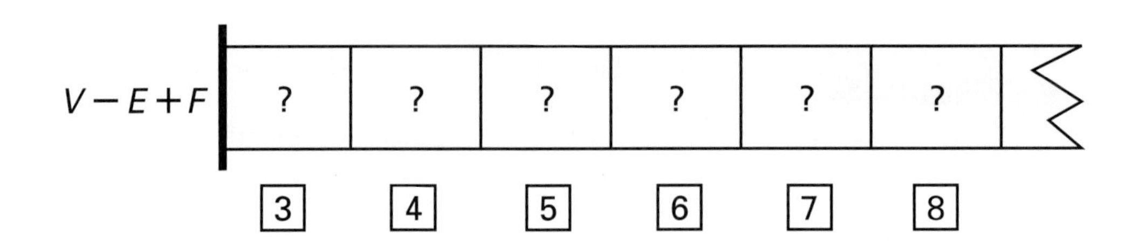

d. What's special about the number strip in part **c**? Explain this special property using the expressions for V, E, and F.

Prisms

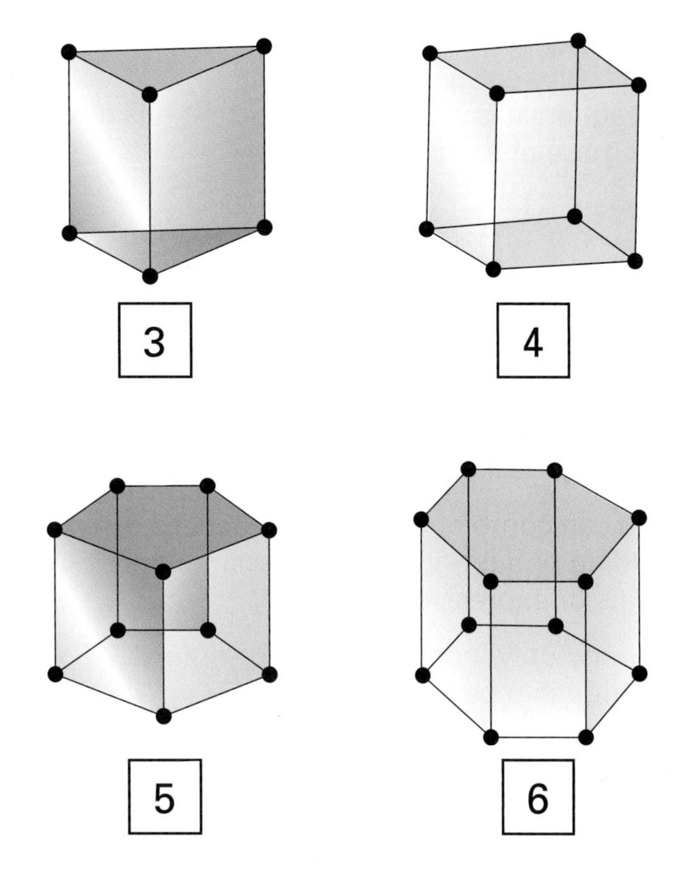

20. a. Make number strips *V*, *E*, and *F* for the sequence of prisms shown above.

b. Write expressions for number strips *V*, *E*, and *F* (expressed in terms of *n*).

c. Use the number strips or the expressions to check that $V - E + F = 2$ for prisms.

The formula $V - E + F = 2$ is called Euler's formula (Euler is pronounced "Oiler"). You have seen this formula in the *Packages and Polygons* unit. The formula works for many polyhedra. For example, an **icosahedron** has 20 faces. For any icosahedron, $V = 12$, $E = 30$, and $F = 20$. Using these values gives $12 - 30 + 20 = 2$.

B Sequences

Summary

A sequence is called arithmetic if it has a constant increase or decrease at each step.

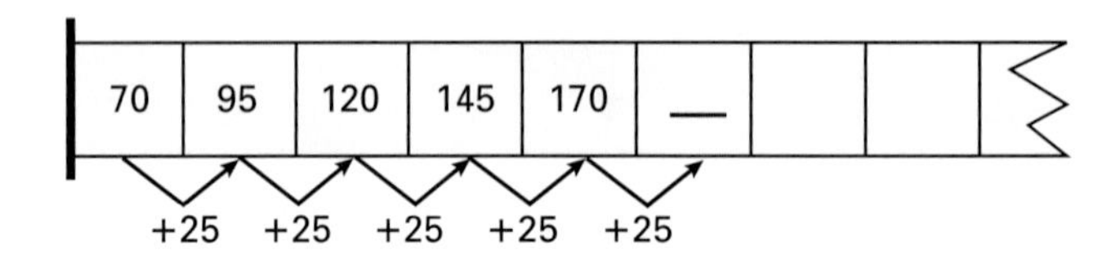

At n steps from the starting number 70, you get the number $70 + 25n$. This expression represents the sequence. Note that n starts at zero.

You can combine number sequences by adding or subtracting them. Adding or subtracting number sequences can be done using number strips or expressions.

For any polyhedron, Euler's formula $V - E + F = 2$ gives a relationship between the numbers of vertices, edges, and faces.

Many other sequences are not arithmetic, for instance the sequence formed by multiplying each term by $\frac{1}{2}$:

$$1, \frac{1}{2}, \frac{1}{4}, \frac{1}{8}.$$

A recursive formula for this sequence is NEXT = CURRENT × $\frac{1}{2}$

Check Your Work

Belinda and Carmen are saving money from part-time jobs after school.

1. a. Belinda currently has $75. She decides to add $5 to her savings each week. Make a number strip that begins with 75 and shows Belinda's total savings every week. What are her savings after n weeks?

b. Carmen currently has $125. Every week she adds twice as much as Belinda does. What are her savings after n weeks?

Look again at the sequence.

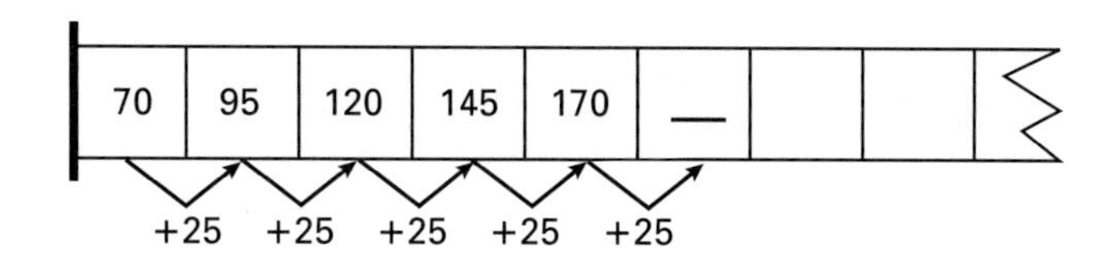

2. a. What is the 15th number in this sequence?

   b. When does the value exceed 1,000 for the first time?

3. a. Make your own arithmetic sequence using fractions.

   b. Write an expression that represents your sequence.

4. **Reflect** If you add two arithmetic sequences, do you always get an arithmetic sequence? Explain why or why not.

A five-sided tower is made by putting a five-sided pyramid on top of a five-sided prism, as shown below.

For this tower:

$V = 11$

$E = 20$

$F = 11$

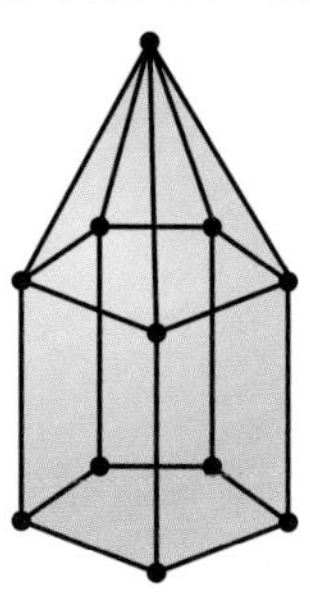

5. a. Does Euler's formula work for a five-sided tower? Explain your answer.

   b. Check to see whether Euler's formula works for an n-sided tower.

For Further Reflection

Can you find a solid for which Euler's formula does not work? If you can, give an example.

Square Numbers

Looking at Squares

The Jacksons want to tile a square patio in their backyard. They bought 200 tiles; each tile measures 30 cm by 30 cm.

1. What are the dimensions of the largest square patio they can make?

After looking at several plans, they decide to arrange the tiles in a more imaginative way. The plan below shows four tiled squares.

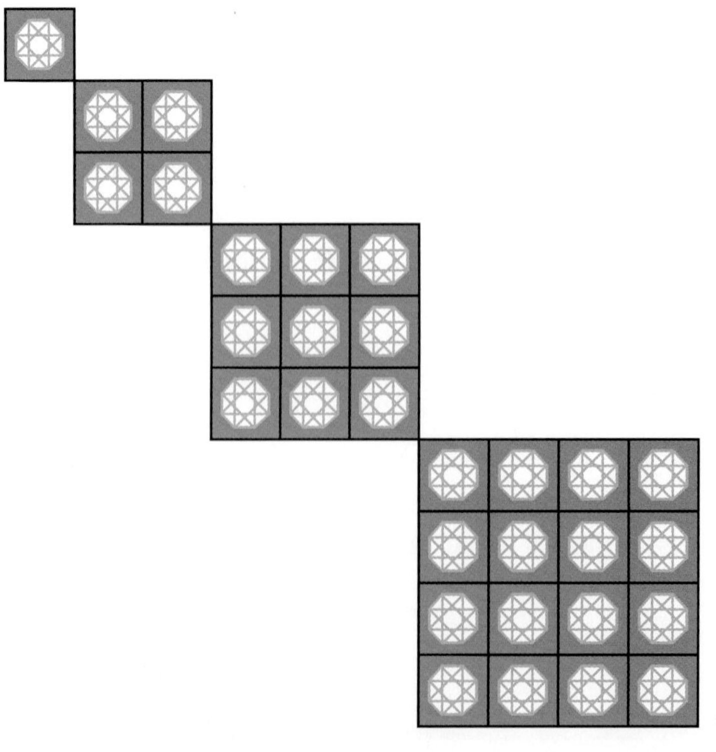

2. What is the largest square the Jacksons can make using this design with 200 tiles available?

To solve problem 2, you may have used the sequence of **square numbers.**

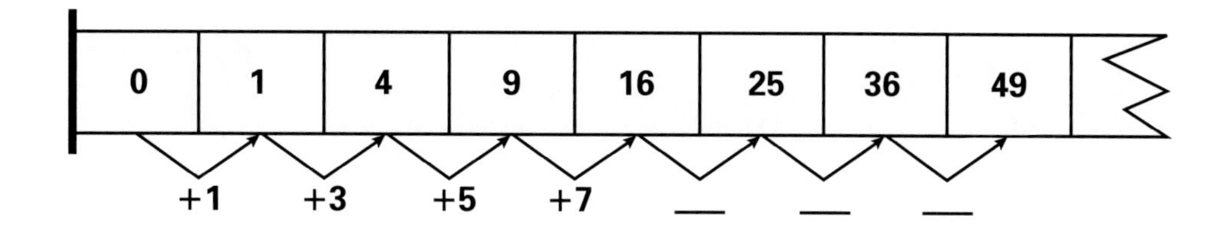

3. a. Why is 0 considered a square number?

b. Describe the increases in the sequence of squares.

c. Is this an arithmetic sequence? Why or why not?

The term "perfect square" becomes clear if you look at dot patterns. You can show 16 as a dot pattern with 4 rows of 4 dots.

4. Use the dot patterns below to describe the increase in the sequence of perfect square numbers.

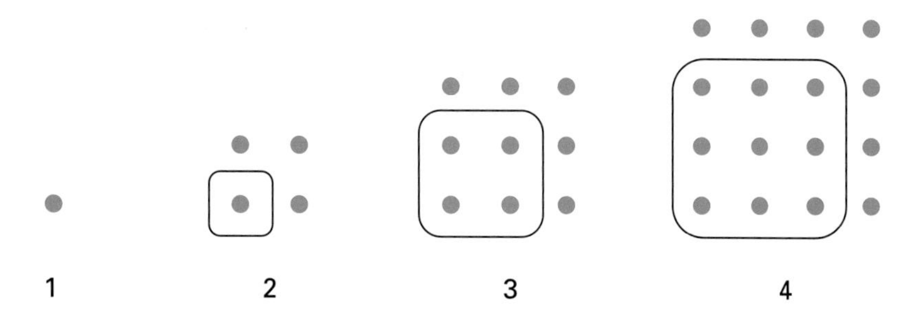

The squares of 20 and 21 are 400 and 441, respectively. This can be written as $20^2 = 400$ and $21^2 = 441$.

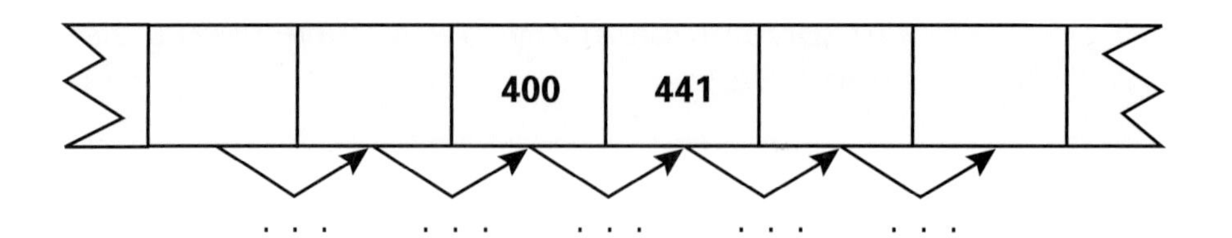

5. **a.** Without multiplying, find the square of 22.
   b. Do the same for the square of 18.

The sequence of square numbers forms several patterns.

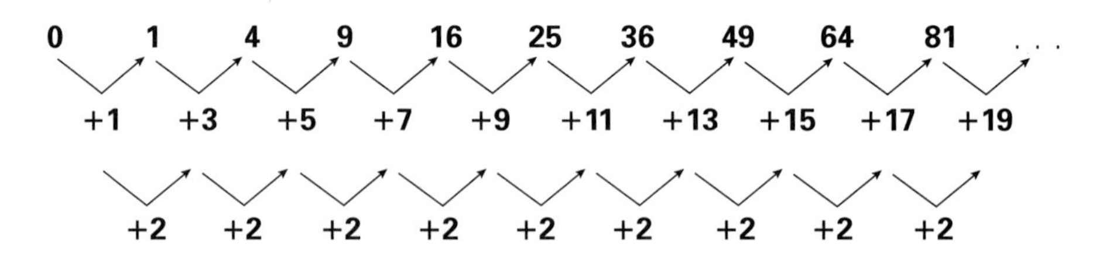

6. **a.** Describe the patterns of the sequence of the square numbers in your own words.
   b. Continue the sequence of squares for six more numbers.

Area Drawings

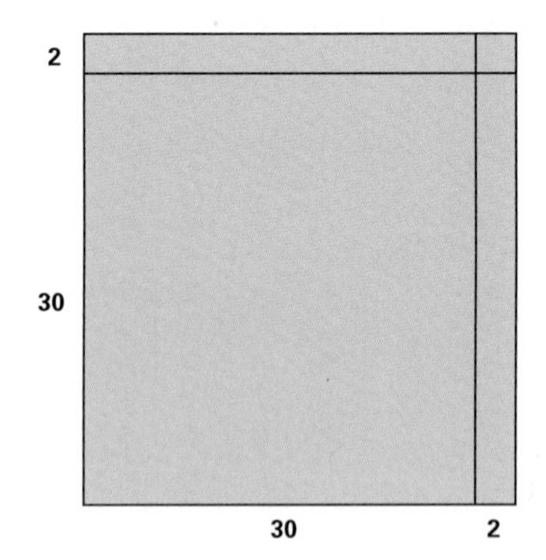

Another way to find the square of a number is by using a diagram such as this one.

7. Explain how to find 32^2 using the diagram.
8. **a.** Make a drawing like the one shown here to help you calculate 43^2.
   b. Do the same for 57^2.

Donald wrote down these calculations:

$$\begin{array}{r} 200^2 = 40{,}000 \\ +\quad 1^2 = \quad\quad 1 \\ \hline 201^2 = 40{,}001 \end{array}$$

9. Do you agree with the calculations above? Explain why or why not.

10. Use the area diagram to show that $(2\frac{1}{2})^2 = 6\frac{1}{4}$.

11. Jackie has a square patio made of 1,444 square tiles. She wants to extend two sides of the patio as shown below. How many extra tiles does she need?

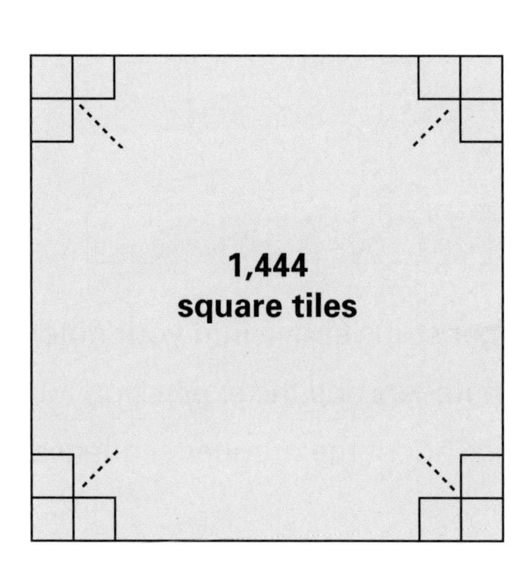

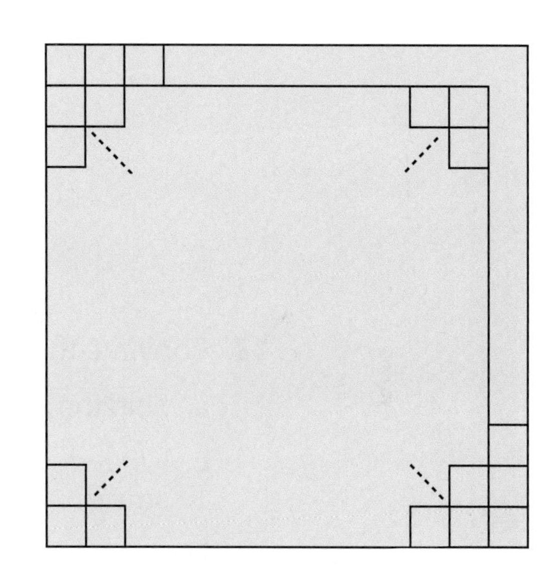

Shifted Strips

The square of a number n can be written as $n \cdot n$ or as n^2. The next number after n can be written as $n + 1$, so the square of $n + 1$ is $(n + 1) \cdot (n + 1)$ or $(n + 1)^2$.

12. How can you write the next number after $n + 1$? How can you write the square of that number?

Larry (the boy who was investigating the odd numbers) decides to add his sequence of odd numbers to the n^2 sequence (where n starts at 0).

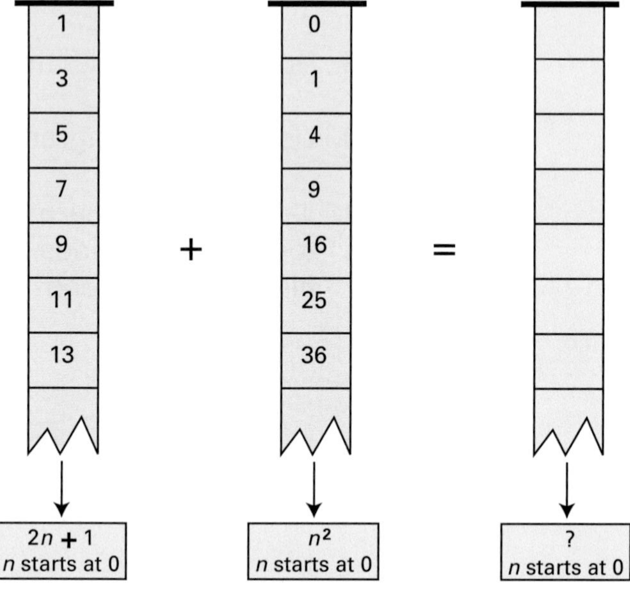

13. Copy the three number strips above into your notebook.

a. Find the missing numbers and the expression for the third strip.

b. What do you notice about the number sequence in the third strip?

14. Larry wrote the expression $(n + 1)^2$ for the third strip. Is he correct? Explain why or why not.

15. If you add the expressions of the first and second number strip, what is the answer?

Larry says, "I have two expressions for the same number strip: $(n + 1)^2$ and $n^2 + 2n + 1$. The two expressions must be equivalent."

16. Copy and use the diagram to explain Larry's statement, $(n + 1)^2 = n^2 + 2n + 1$, in a different way.

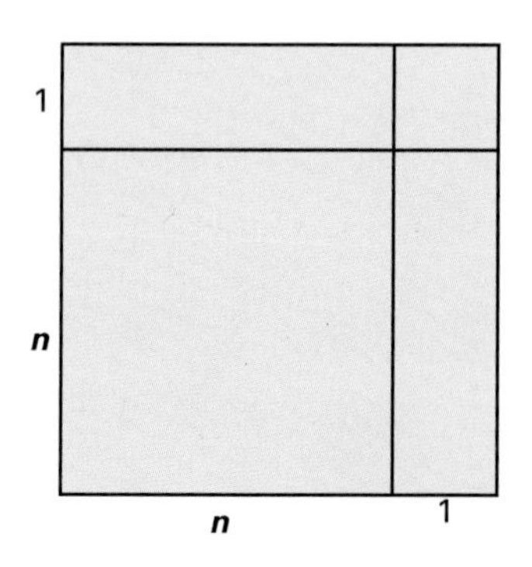

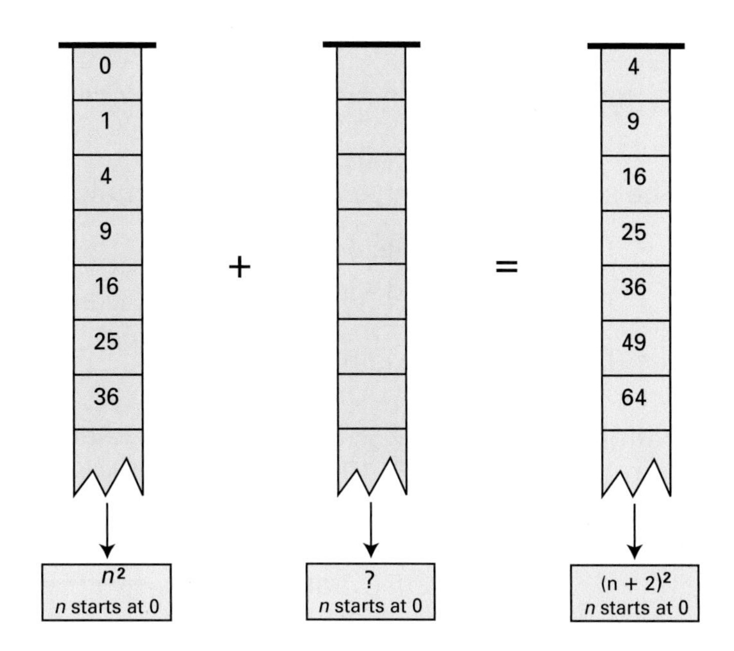

The $(n + 1)^2$ sequence starts one step later than the n^2 sequence. Larry wonders how to get a sequence that starts two steps later.

17. a. Copy and complete the number strip in the middle. Assume that n starts at 0 for each strip.

b. What expression describes the second strip? (Be sure n starts at 0 with your expression.)

c. Use the expressions from the first two strips to write an equivalent expression for $(n + 2)^2$.

d. Use an area diagram to explain your answer in part **c.**

You can keep starting the sequence later by adding one more to n each time, as shown below.

Sequence	**Formula, where *n* starts at 0 in each case**
0, 1, 4, 9, 16,	$n^2 = n^2$
1, 4, 9, 16,	$(n + 1)^2 = n^2 + 2n + 1$
4, 9, 16,	$(n + 2)^2 = n^2 + 4n + 4$
9, 16,	$(n + 3)^2 = n^2 + 6n + 9$

18. By looking at patterns, predict what the equivalent expression for $(n + 4)^2$ would be. Where does the sequence start?

19. Find an equivalent expression for $(2n + 1)^2$. Explain your method and how you know it is correct.

Square Numbers

Summary

The square of a number is the number multiplied by itself. For example:

- The square of 4 is $4 \times 4 = 4^2 = 16$.
- The square of $3\frac{1}{2}$ is $3\frac{1}{2} \times 3\frac{1}{2} = (3\frac{1}{2})^2 = 12\frac{1}{4}$.
- The square of n is $n \cdot n = n^2$.

Numbers like 4, 9, $12\frac{1}{4}$, 36, 10,000 are called square numbers, and numbers like 4, 9, 36, 10,000 are perfect squares.

The same number strip can represent equivalent expressions.

$(n + 1)^2$ and $n^2 + 2n + 1$ are both represented by 1, 4, 9, 16, 25, …, where n starts at zero.

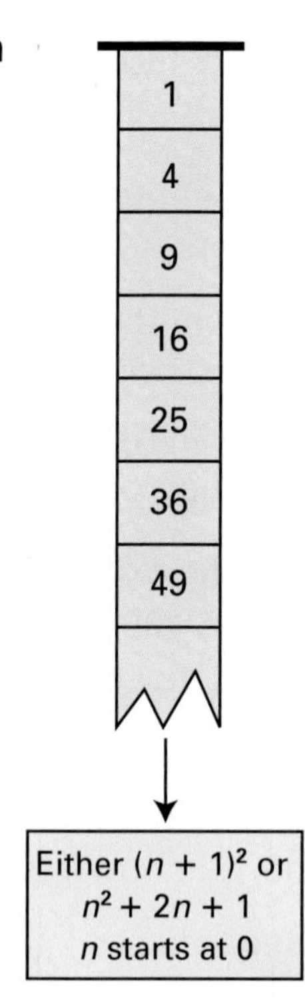

An area diagram can also be used to show that $(n + 1)^2 = n^2 + 2n + 1$.

1	$1 \times n$	1×1
n	$n \times n$	$n \times 1$
	n	1

Check Your Work

1. What is the largest square patio you can make with 68 square tiles? You may use only whole tiles.

2. Write down two square numbers between 30 and 40.

If you start the n^2 sequence three steps later, you get the $(n + 3)^2$ sequence.

3. Show that $(n + 3)^2$ is the same as the expression $n^2 + 6n + 9$ by using an area diagram.

4. Write an expression that is equivalent to $(n + 10)^2$.

Here is a sequence of squares that can be extended as far as you wish.

0	4	16	36	64	100	144	....

5. Use regularities in the pattern to find the next three numbers of the sequence. (Hint: Look again at problem 6.)

For Further Reflection

Sully says that the square of a number is the same as the number times two. Is he ever right? Explain why or why not.

D

Triangles and Triangular Numbers

Tessellations and Tiles

Here is a collection of tiles. You may have seen these shapes before in the *Packages and Polygons* unit.

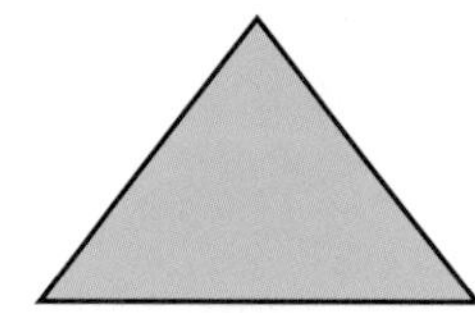
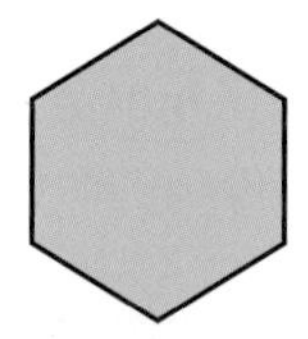
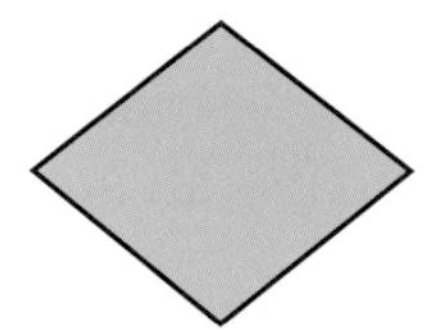
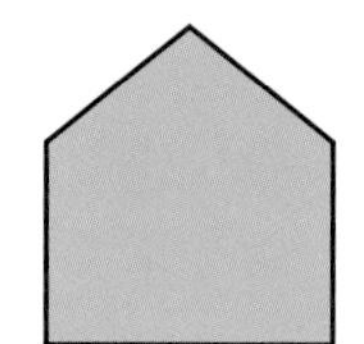

1. Name the shape of each tile above.

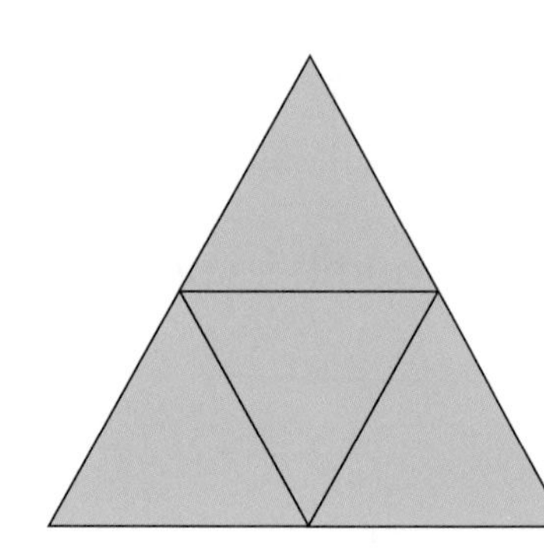

In the *It's All the Same* unit, you saw that you can tessellate or cover the plane using any triangle. With a triangular tile, you can also tessellate a triangle that is exactly the same shape as the basic tile. On the left, an equilateral triangle tessellates a larger equilateral triangle.

If you alternate purple and white tiles, you can create triangles with interesting patterns.

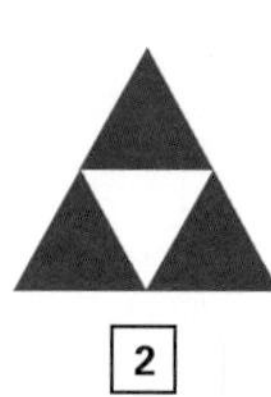

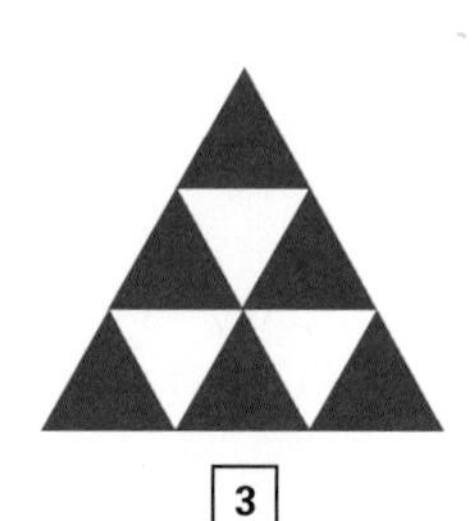

2. a. Find the total number of tiles used for each of the triangles.

b. How many tiles do you need to add a row to the base of the fourth triangle to get the fifth pattern?

3. Find the total number of tiles needed to make the tenth triangle in the sequence.

The number of tiles in each triangle in the sequence is shown on the number strip.

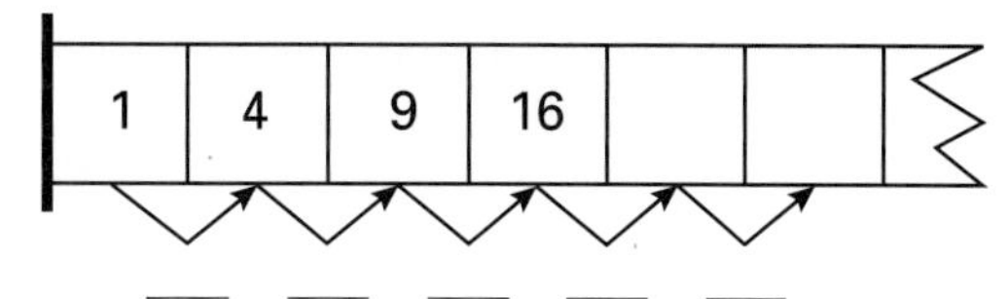

4. a. What is the 30th number in this sequence?

   b. Use n to write an expression for this number strip. Where does n start?

Janet is building a triangle table. She wants to completely cover the table with triangle tiles. A newspaper is covering most of the table so that only one row of triangles can be seen.

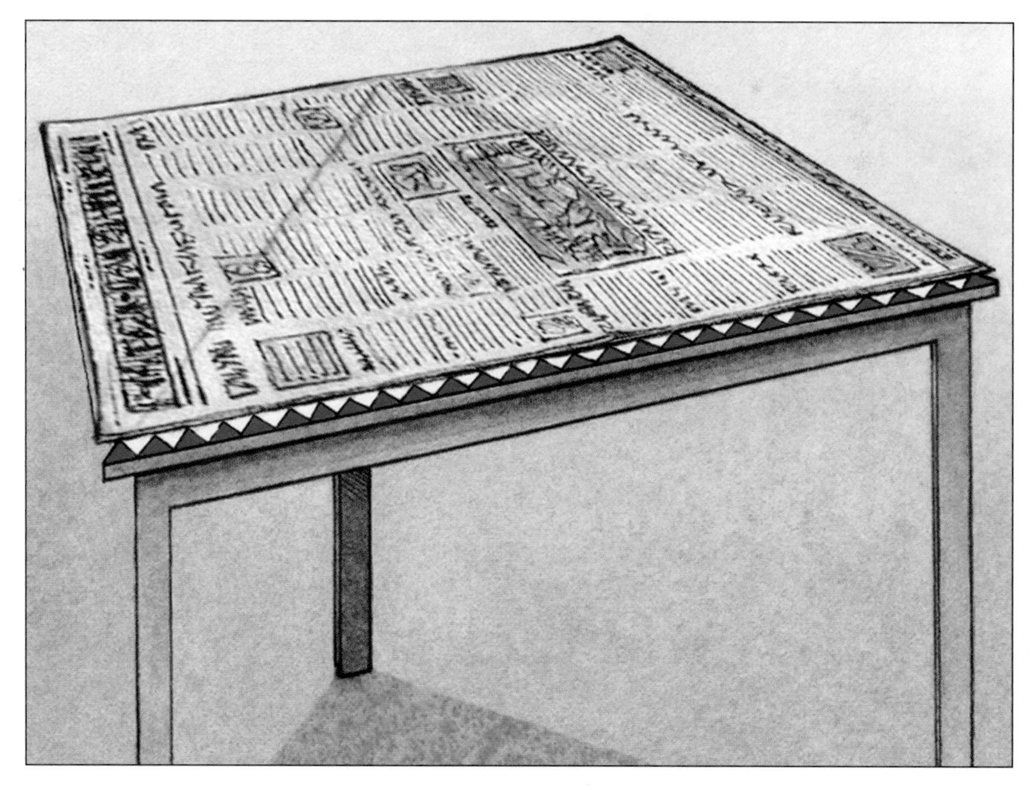

5. a. How many rows of triangles cover the table?

   b. Find the total number of tiles that are hidden by the newspaper.

Activity

Triangular Patterns

As you found on the previous page, you can use a simple rule to calculate the number of tiles in a triangular **tessellation**. If there are n tiles along the base of the tessellation, then the total number of tiles is equal to n^2.

Study the triangular tessellations to see one explanation of why the rule stated above works.

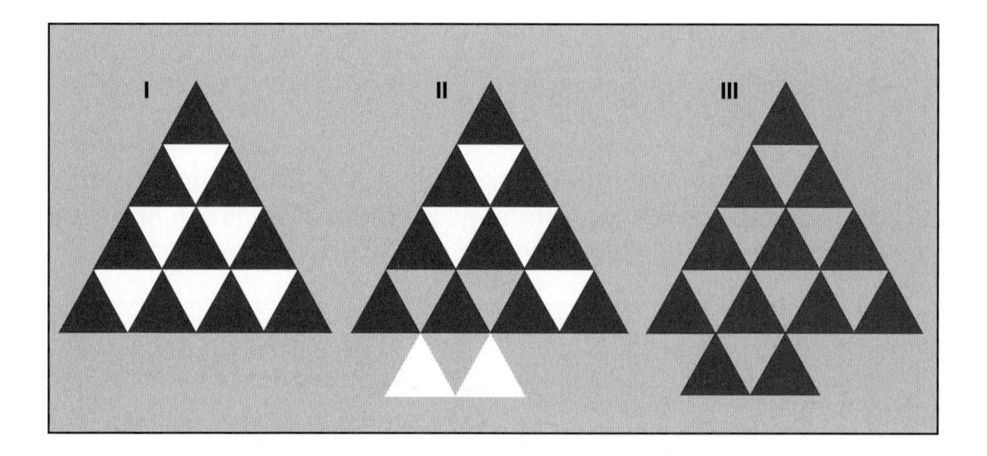

Use **Student Activity Sheet 2** to answer problem 6.

6. a. Starting with Figure I, reflect every white triangle over the base of the triangle (Figure II). Color each reflected triangle red (Figure III).

b. Explain why the total number of tiles in Figure I is equal to the number of red tiles in the finished version of Figure III.

c. The rule states that the total number of tiles in Figure I is equal to 4^2. Explain this using the finished version of Figure III.

d. Verify the rule for a triangle with five rows ($n = 5$).

7. **a.** Determine how many white triangles are in the tessellation above without counting each one. Explain your method.

   b. How many red triangles are there?

Triangular Numbers

In the previous pages, you studied patterns with triangles (as in the first picture below).

The Greek mathematician Nikomachos, who lived around 100 A.D., studied triangular dot patterns. The numbers, 1, 3, 6, 10,.... are called **triangular numbers.**

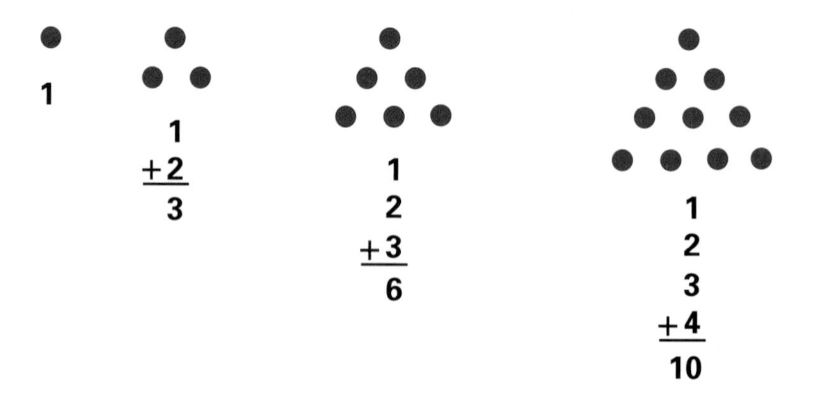

8. a. Describe any connections that you see between the triangular tessellations, the dot patterns, and the triangular numbers.

b. What regularities do you see in each of the patterns above?

9. The 30th triangular number is 465. Using this fact, what is the 31st triangular number? What is the 29th?

Rectangular Numbers

Nikomachos was interested in finding a direct formula for the triangular numbers.

As part of his investigation, he used a pattern he called the **rectangular numbers**.

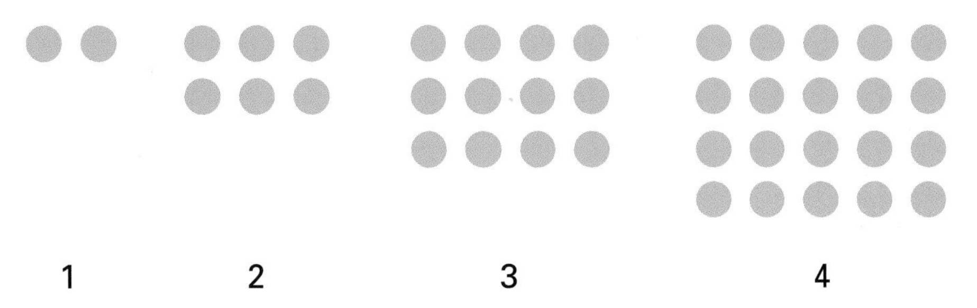

The rectangular numbers are 2, 6, 12, 20, and so on.

10. a. Find the next three rectangular numbers.

b. Is 132 a rectangular number? Explain why or why not.

11. Ann wrote the formula $R = n(n + 1)$ for rectangular numbers. Barbara wrote a different formula: $R = n^2 + n$. Use dot patterns to explain both Ann's formula and Barbara's formula.

12. a. Make a drawing to demonstrate that each rectangular number is double a triangular number.

b. Nikomachos found a direct formula for the triangular numbers:

the nth triangular number $= \frac{1}{2}n(n + 1)$,

where n starts at 1.

Explain this formula.

13. In problem 9, you found the 29th, 30th, and 31st triangular numbers. Use Nikomachos's formula to check those values.

Activity

A Wall of Cans

14. How many cans will fit in a triangular display against this wall?

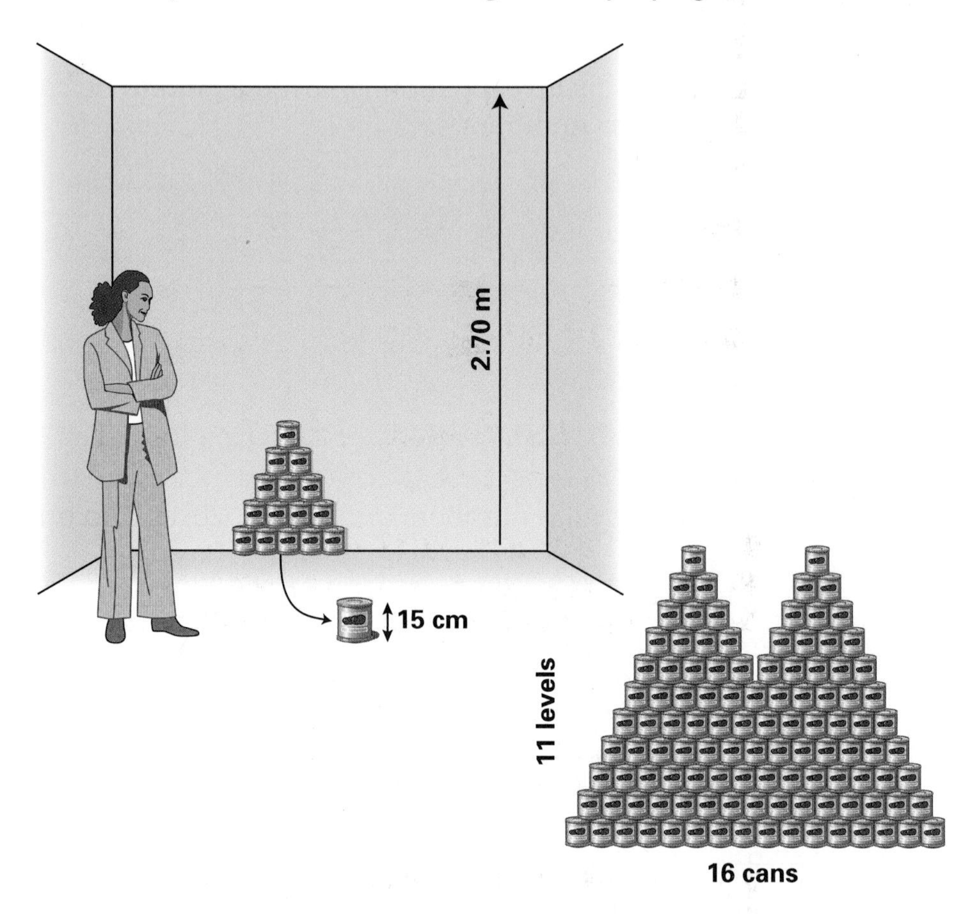

The manager of a store wants to try a new way to stack cans. She thinks a camel shape would be more eye-catching, but she is not sure how many cans would be needed to build that shape.

Use **Student Activity Sheet 3** to help you solve problem 15.

15. a. If the display is to be 16 cans wide with 17 levels, study the drawing and find the number of cans needed for the display.

b. Write the steps in your calculation. Make sure your steps are clear so that you could follow those same steps for a different number of cans in the bottom layer and for a different number of levels.

c. Design a new display. Draw the shape for your own arrangement of cans. Include the important measurements and predict how many cans will be required.

The Ping-Pong Competition

The Jefferson Middle School Student Council wants to organize a ping-pong competition. Everyone who enters the competition will play against everyone else.

The Student Council wants to know how many games will be played. You can use patterns to find the answer for them.

Number of Players	Graph	Number of Games
2		1
3		3
4		6

16. Copy the table into your notebook. Continue the table for five players and for six players.

17. For six players, how many lines are drawn from each vertex?

18. Look at the table to find a pattern. Use your pattern to predict the number of games for seven players and for eight players. Check your answers by extending the table in your notebook.

19. If 50 games are the most that can be played, how many participants can compete?

20. Write a formula that you can use to compute the number of games for any number of players.

D Triangles and Triangular Numbers

Summary

In this section, you found a simple rule: The total number of small triangles that tessellate a larger triangle with n rows equals n^2.

You also studied two types of dot patterns.

Rectangular Pattern

Triangular Pattern

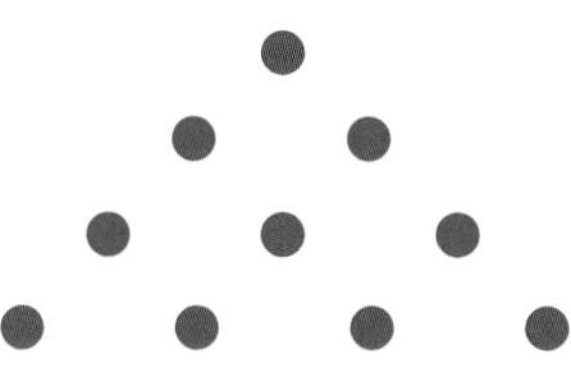

The rectangular numbers are 2, 6, 12, 20, and so on.
The nth rectangular number is $n(n + 1)$, where n starts at 1.

The triangular numbers are 1, 3, 6, 10, and so on. When you look at the dot patterns, each rectangular pattern can be divided into two triangular patterns, so the nth triangular number is $\frac{1}{2}\, n(n + 1)$.

Check Your Work

1. **a.** Describe the pattern in the tiles in the tessellation shown in the Summary.

   b. Explain how you can find the total number of red tiles and of white tiles without counting them.

2. a. Suppose you have a stack of pipes, like the one shown on the right, with 5 pipes on the bottom and 1 pipe on the top. Compute the number of pipes in the stack. Use some method other than counting each one.

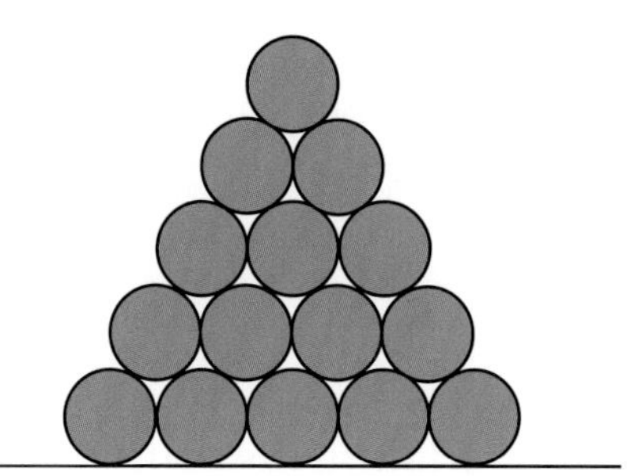

b. Compute the number of pipes in a stack that has 25 pipes on the bottom and 1 pipe on the top. Use some method other than counting each one.

c. Design and solve your own pipe problem.

3. If you started counting the dots from the top of the triangle, going down by rows, how many dots in total have you counted when you reach the circled dot? Use what you know about triangular numbers to answer this problem.

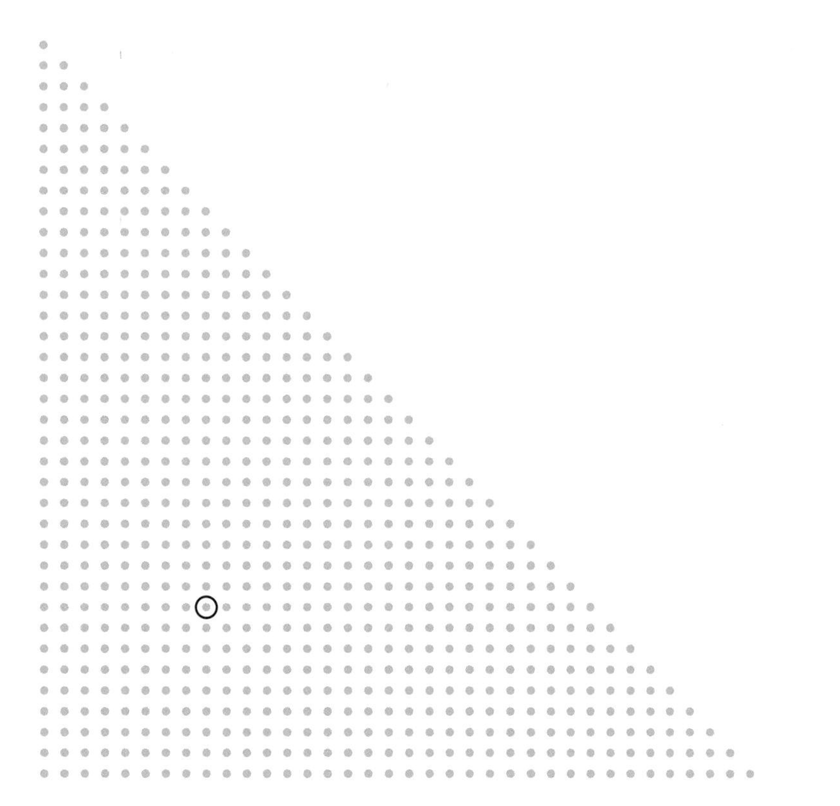

Triangles and Triangular Numbers

In the MiC Tennis Tournament, everyone who enters the competition will play against everyone else. Suppose 12 players from Rydell Middle School will play.

4. How many games do the 12 players play in total?

Before the tournament starts, each participant shakes hands with all competitors.

5. How many handshakes are given in total?

For Further Reflection

Find a situation that has the same mathematical content as the ping-pong tournament and the handshake problem.

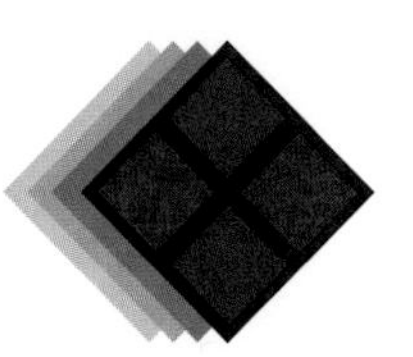

Additional Practice

Section A Patterns

1. Write the first five numbers in each of the sequences described by the following expressions. For each expression, n starts at zero.

   a. $2n + 3$

   b. $15n - 10$

   c. $\frac{1}{2}n + \frac{1}{2}$

2. Make your own expression and write the first five numbers of the sequence represented by your expression. Make sure n starts at zero.

3. a. Write a NEXT-CURRENT formula for the dot pattern shown below.

   b. Describe the dot pattern with a direct formula $D =$ ....

Dot Pattern:

Pattern Number: 2 3 4

4. a. Make a number strip for the formula

NEXT number = CURRENT number + 4, with starting number 17

   b. Write an expression that represents the sequence.

Section B Sequences

Joey and Alice collect old magazines for their school. Joey has currently collected 24 magazines. Each week he gets three more old magazines.

1. Make a number strip that begins with 24 and gives the number of magazines Joey has at the end of each week.

2. How many magazines will Joey have after n weeks?

Alice currently has 39 magazines. Each week she collects two magazines.

3. How many magazines will Alice have after n weeks?

4. How many magazines will Joey and Alice collect together after n weeks?

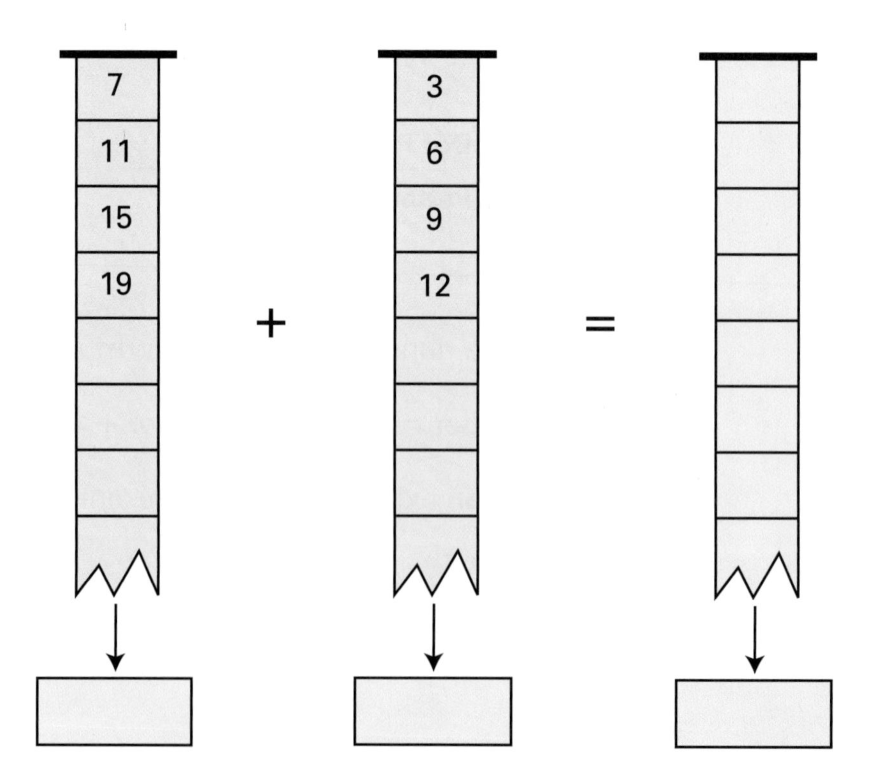

5. Copy and complete the missing parts of the three number strips shown above. Let n start at zero for all three number strips.

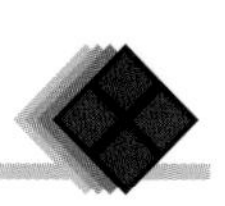

6. Copy and complete the missing parts of the number strip subtraction shown below. Let n start at zero for all three number strips.

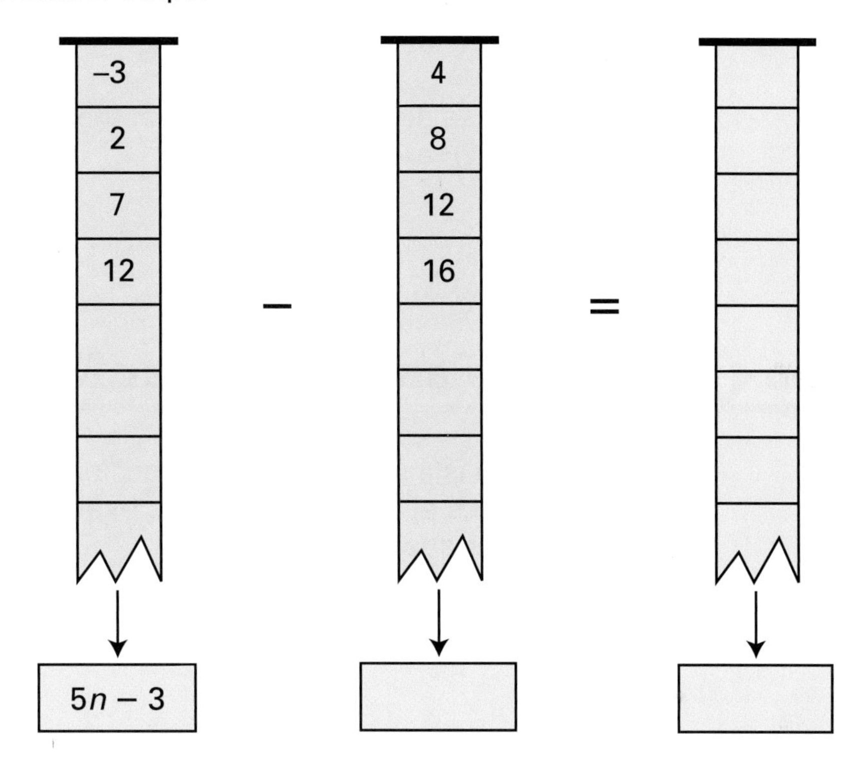

Section C Square Numbers

Here is a sequence of four tile patterns. P stands for the pattern number.

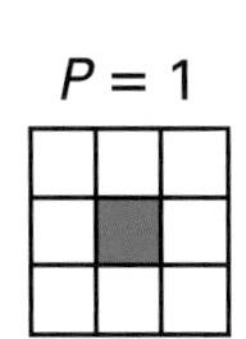

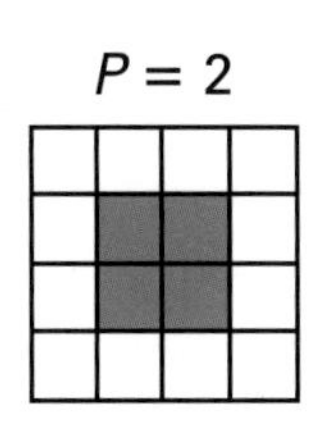

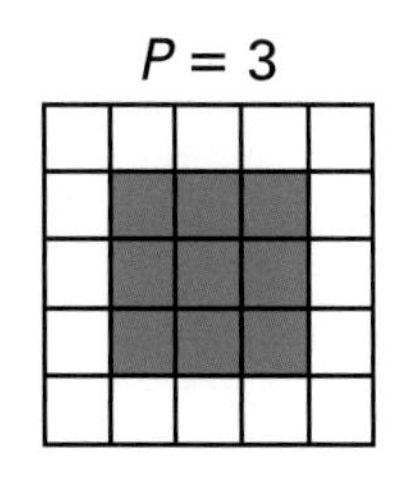

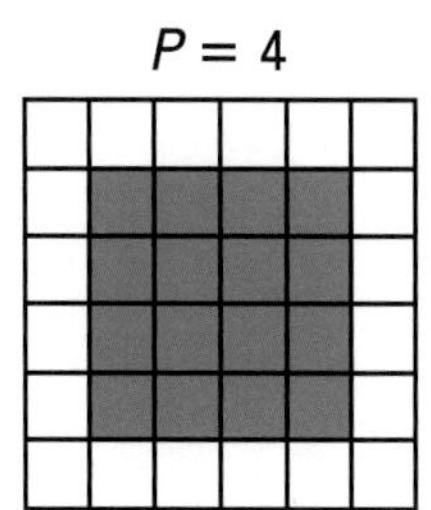

1. Write a direct formula to calculate the number of green tiles needed for each pattern number (P). Note that P starts at one in this problem!

2. Explain using the tile patterns above that the formula for the number of white tiles for pattern number P is

number of white tiles = $4 \times (P + 1)$

3. What is the formula for the total number of tiles?

4. Explain how the formulas you found in problems 1, 2, and 3 are related.

5. Use an area diagram to show that the expressions $n^2 + 4n + 4$ and $(n + 2)^2$ are equivalent.

Section D Triangles and Triangular Numbers

A tetrahedron is a regular three-dimensional shape with four equal faces. The faces each have the shape of an equilateral triangle. A picture of a tetrahedron is shown.

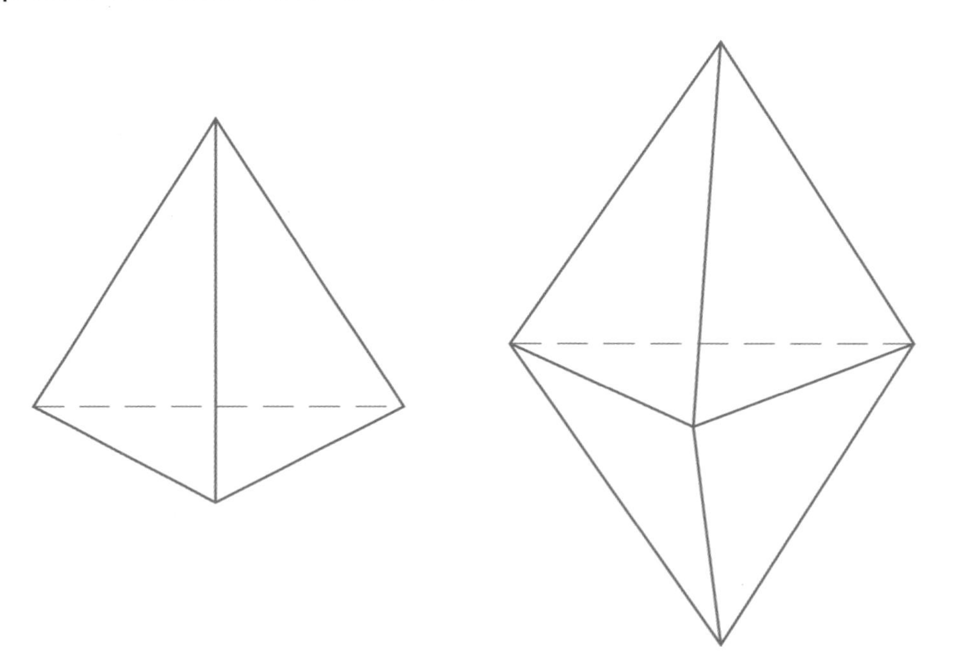

Anton has started to cover each face with blue and white triangular tiles.

He can fit 11 blue tiles along each edge.

1. How many tiles (blue and white) does Anton need to completely cover the tetrahedron?

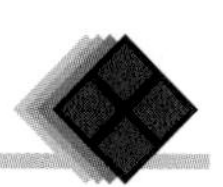

You can paste two tetrahedra together as shown in the figure.

2. How many tiles are needed to completely cover this new shape?

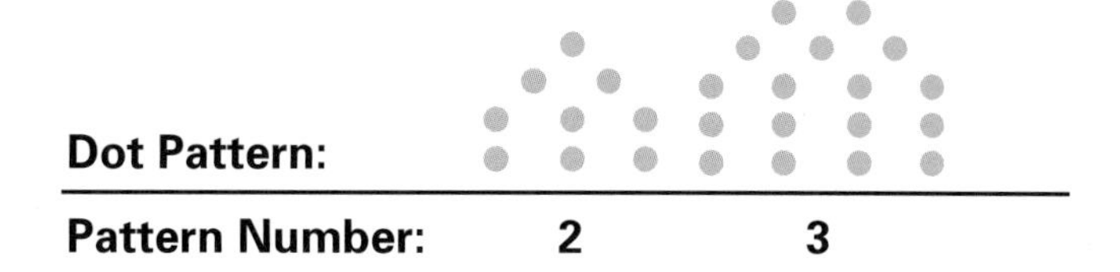

3. Study the dot pattern above and draw the pattern for $n = 4$.

You can split each pattern in a triangle and a rectangle so that the base of the triangle has as many dots as the right side of the rectangle. The following is a sketch of how the shape can be split into a triangle and a rectangle.

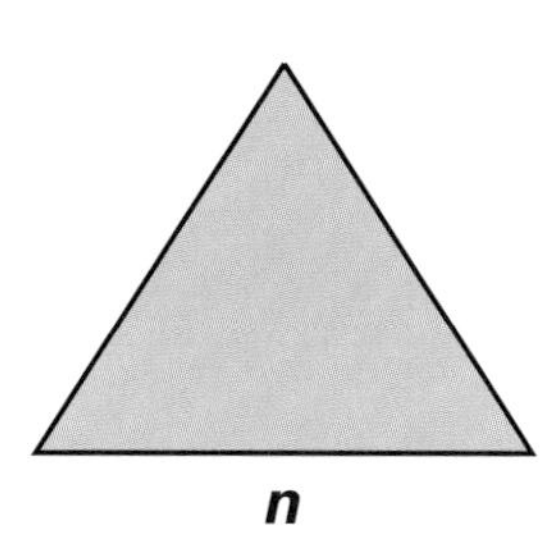

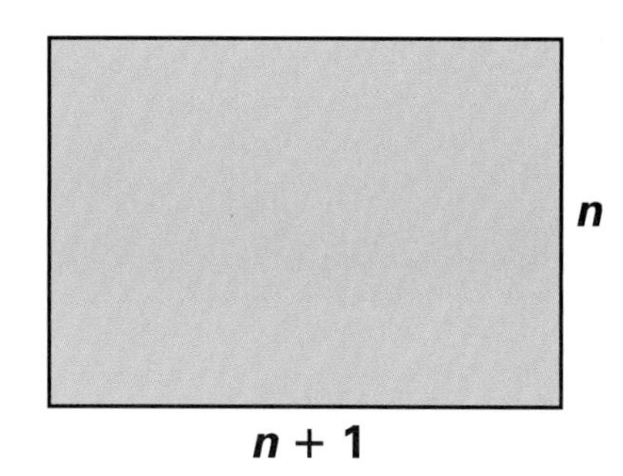

4. a. Draw the triangle for the pattern $n = 5$.

   b. Draw the rectangle for the pattern $n = 5$.

   c. How many dots are needed for the pattern $n = 5$?

5. Write an expression for the number of dots for pattern n. Use the sketch above.

Answers to Check Your Work

Section A Patterns

1. James's formula is correct. One way to show this is to point out that each pattern has n rows of three dots, plus one on top: $3n + 1$.

2. a. Yes, both formulas are correct. Your explanation may differ from the ones presented here. If that is the case, discuss it with a classmate. Sample explanations:

   - If you fill in numbers 0, 1, 2, 3, etc. in both formulas, you get the same dot pattern.
   - $2(n + 1) - 1 = 2n + 2 - 1 = 2n + 1$.

   b. A table may help you find the answer for this problem.

Pattern Number	Number of Dots
1	1
2	3
3	5
4	7
5	9
6	11

David's formula would change into $W = 2n - 1$ or $W = 2(n+1)-3$.

3. a. You may have different patterns. Sample dot pattern:

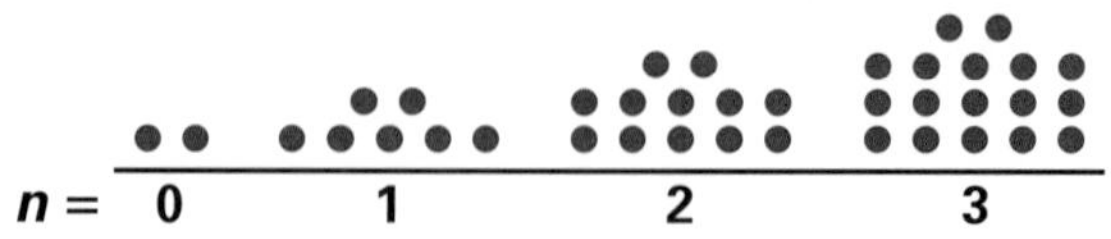

$n =$ 0 1 2 3

   This is a good way to record pattern and pattern number.

   b. START NUMBER = 2
   NEXT = CURRENT + 5

4. a. There are many possible patterns. Discuss your pattern with a classmate. Here is a sample pattern.

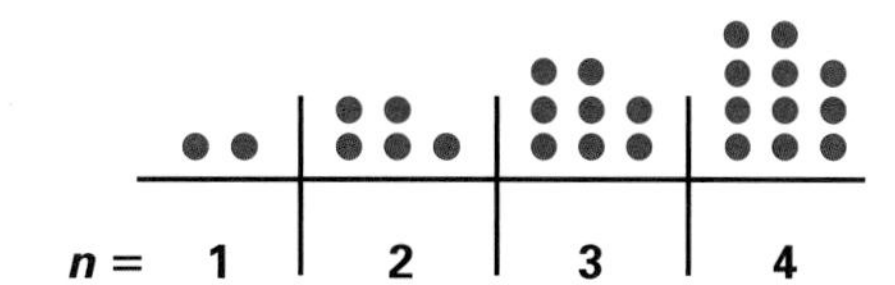

b. Make sure your direct formula and your NEXT-CURRENT formula correspond with your sequence. Fill in $n = 3, 4, 5$, etc. to check.

A direct formula for the sample sequence of 4a is: $D = 3n - 1$; n starts at 1. (D represents the number of dots.)

A NEXT-CURRENT formula is

NEXT = CURRENT + 3; START number is 2

5. An advantage of a NEXT-CURRENT formula is that it is often easier to make. You only have to look at the start number and the increase or decrease.

A disadvantage of a NEXT-CURRENT formula is that it does not immediately give you the number of dots for any pattern number in the sequence. You have to generate all of the elements before the one in which you are interested in order to know its value.

If you found other advantages or disadvantages, discuss those in class.

Section B Sequences

1. a. After n weeks, Belinda has $75 + 5n$.

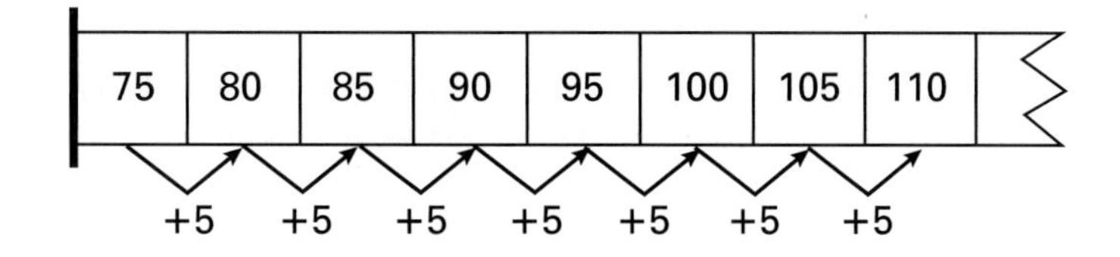

b. $125 + 10n$. You might reason in one of the following ways:

- I can make an expression by seeing that 10 is double 5, and n still stands for the number of weeks.
- I made a strip first and then found an expression for that strip.

2. **a.** The 15th number is 420. The first number is 70, with n = 0. The 15th number will be n = 14, so
   $70 + 25 \times 14 = 420$.

   b. The value exceeds 1,000 on the 39th number (when *n* is 38 or larger). Strategies will vary. Sample strategies:

   - If the value of $70 + 25n$ must exceed 1,000, $25n$ must exceed 930. Therefore, *n* must be 38.
   - $930 \div 25 = 37.2$, which can be rounded to 38.
   - Multiply 25 by different numbers until the answer exceeds 930 (accounting for the additional 70).

   Continue to fill out the table until the answer exceeds 930.

3. **a.** Compare your sequence with that of a classmate. Let him or her check whether your expression fits. A sample arithmetic sequence using fractions is: 10, $7\frac{1}{2}$, 5, $2\frac{1}{2}$, 0, $-2\frac{1}{2}$, -5, $-7\frac{1}{2}$, etc. The constant *decrease* in this sample sequence is $2\frac{1}{2}$.

   b. An expression that represents the sample sequence of 3a is $10 - 2\frac{1}{2}n$, *n* starts at zero.

4. Yes, when you add two arithmetic sequences together, you add the starting points ,and you add the two changes. That means the new sequence will start at the sum of the two starts and will change by the sum of the two changes.

5. **a.** Yes. Euler's formula is $V - E + F = 2$; substituting the given values you get $11 - 20 + 11 = 2$.

   b. Yes. For an *n*-sided tower,

   $V = 2n + 1$

   $E = 4n$

   $F = 2n + 1$

   $$\begin{aligned} V - E + F &= (2n + 1) - 4n + (2n + 1) \\ &= 2n + 2n - 4n + 2 \\ &= 2 \end{aligned}$$

Section C Square Numbers

1. A square patio of 8 × 8 = 64 tiles, so 4 tiles are left. If you answered a square patio of 17 tiles in length and width, you only placed your squares at the perimeter of the patio and the patio itself is filled with sand.

2. 36 is a square number because 6 × 6 = 36. It is the only perfect square between 30 and 40.

$30\frac{1}{4}$ is a square number because $5\frac{1}{2} \times 5\frac{1}{2} = 30\frac{1}{4}$.

Note that square numbers do not have to be whole numbers!

3.

	n	3
3	$3n$	9
n	n^2	$3n$

4. $n^2 + 20n + 100$. You can find this expression by looking at number strips, by drawing an area diagram as shown below, or possibly by doing symbol manipulation.

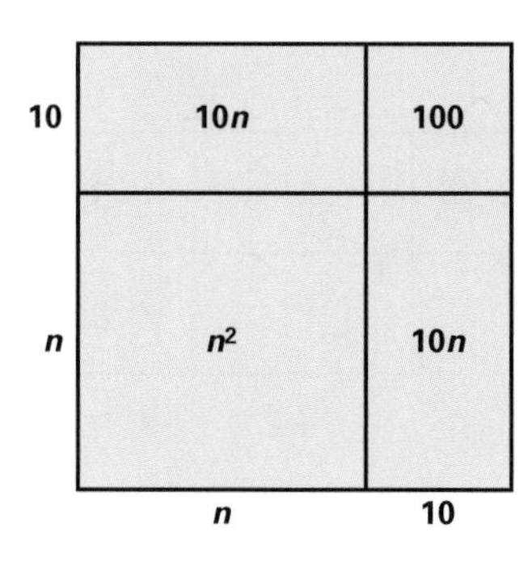

5. The next three numbers of the sequence are 196, 256, and 324. Look at the regularities in the sequence shown below.

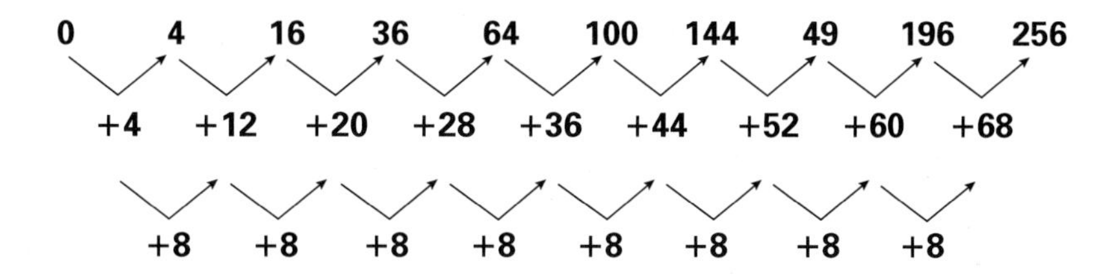

Section D Triangles and Triangular Numbers

1. a. The number of red tiles and the number of white tiles in each row increases according to the pattern in the triangular numbers. The pattern in the red tiles starts with the number 1, and the pattern for the number of white tiles starts with the number 0. You know that because:

- there is a rule that if there are n tiles along the base of a triangular tessellation, then the total number of tiles is equal to n^2. The total number of red and white tiles is 49.

b. You could find the total number of tiles in different ways. For example, look at the pattern for the white tiles:

Row Number from Base	1	2	3	4	5	6	7
Number of White Tiles	6	5	4	3	2	1	0

The total number of white tiles is 6 + 5 + 4 + 3 + 2 + 1 = 21.

The total number of red tiles is 49 − 21 = 28.

- You can look at the pattern of the total number of red triangles after each row.

Row Number from Top	1	2	3	4	5	6	7
Total Red Tiles	1	3	6	10	15	21	28

+2 +3 +4 +5 +6 +7

+1 +1 +1 +1 +1

- The total number of red tiles is 28, so the number of white tiles would be 49 − 28 = 21.

2. a. 15 pipes.

b. 325 pipes. Sample strategies:

Using Nikomachos's formula to find the 25th triangular number:

$= \frac{1}{2} \times 25 \times (25 + 1)$

$= \frac{1}{2} \times 25 \times 26$

$= 325$

By adding the first and last rows, and then the second and next-to-last rows, and so on, you get 12 groups of 26 plus the 13 in the middle: $12 \times 26 + 13 = 325$.

c. Problems will vary. Sample problem:

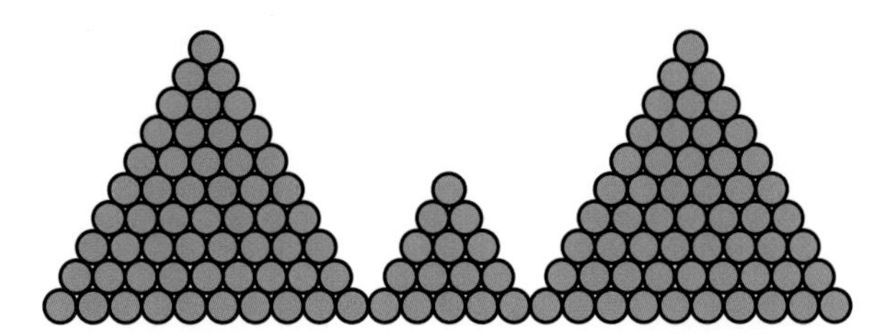

The left and right triangles are identical and have 10 pipes in the base; the middle triangle has a base of five pipes.

Using Nikomachos's formula:

$2 \times \frac{1}{2}(10 \times 11) + \frac{1}{2}(5 \times 6)$

$= 110 + 15$

$= 125.$

3. 387. Sample strategy:

You may think about this problem in different ways. One way is to count the number of dots in the row above the circled dot, which is 27. The 27th triangular number is $\frac{1}{2}(27 \times 28) = 378$. Adding on the nine dots in the row with the circled dot, you get $378 + 9 = 387$.

4. If you need help with this problem, look at the ping-pong competition.

The number of games is $\frac{1}{2} \times 12 \times 11 = 66$.

5. The number of handshakes is the same as the number of games, 66.